URBAN SETTLEMENT *and* LAND USE

D0533187

access to geography

URBAN SETTLEMENT *and* LAND USE

Michael Hill

HODDER
EDUCATION
PART OF HACHETTE LIVRE UK

For nephews Mark and Andrew.

The publishers would like to thank the following individuals, institutions and companies for permission to reproduce copyright illustrations in this book: Columbia/Tri-Star/The Kobal Collection, page 129; © GMPTE, page 95; Taylor and Francis Books Ltd (Routledge), page 33. All other photos are by the author.

The publishers would also like to thank the following for permission to reproduce material in this book: David Fulton Publishers (www.fultonpublishers.co.uk) for an extract from *Cities in Space: City as Place* by D Herbert and C Thomas (1997), page 89.

Every effort has been made to trace and acknowledge ownership of copyright. The publishers will be glad to make suitable arrangements with any copyright holders whom it has not been possible to contact.

Although every effort has been made to ensure that website addresses are correct at time of going to press, Hodder Murray cannot be held responsible for the content of any website mentioned in this book. It is sometimes possible to find a relocated web page by typing in the address of the home page for a website in the URL window of your browser.

Hodder Headline's policy is to use papers that are natural, renewable and recyclable products and made from wood grown in sustainable forests. The logging and manufacturing processes are expected to conform to the environmental regulations of the country of origin.

Orders: please contact Bookpoint Ltd, 130 Milton Park, Abingdon, Oxon OX14 4SB. Telephone: (44) 01235 827720. Fax: (44) 01235 400454. Lines are open 9.00–5.00, Monday to Saturday, with a 24-hour message answering service. Visit our website at www.hoddereducation.co.uk

© Michael Hill
First published in 2005 by
Hodder Education,
part of Hachette Livre UK,
338 Euston Road
London NW1 3BH

Impression number 10 9 8 7 6 5 4 3
Year   2010 2009 2008

Cover photo Giorgio de Chirico, 'The Enigma of a Day', 1914 © DACS 2005. Photo SCALA, Florence/New York Museum of Modern Art.
Typeset in 10/12pt Baskerville and produced by Gray Publishing, Tunbridge Wells
Printed in Malta

A catalogue record for this title is available from the British Library

ISBN: 978 0340 883 457

Contents

1 An Introduction to Urban Settlement and Land Use

'Meanwhile, at social Industry's command
How vast an increase. From the germ
Of some poor hamlet, rapidly produced
Here a huge town, continuous and compact
Hiding the face of earth for leagues – and there,
Where no habitation stood before
Abodes of men irregularly massed
Like trees in forests – spread through spacious tracts,
O'er which the smoke of unremitting fires
Hangs permanent, and plentiful as wreaths
Of vapour glittering in the morning sun.'

William Wordsworth *The Excursion*

The human population is living in an ever-increasingly urbanised world. At the beginning of the twentieth century, an estimated 10% of the human population lived in cities. At some time around 20 years into the new millennium, this figure will have grown to just over 50%. By then, the majority of the world's people will no longer be rural dwellers working on the land, but will be living in towns and cities and engaged in manufacturing or service industries. Britain was the first country to undergo urbanisation of the modern, industrial type, during the eighteenth and early nineteenth centuries and it was in reaction to this that Wordsworth wrote the lines quoted above in 1814. The process of urbanisation has since then diffused throughout the world, first to other parts of Europe, then to North America and Japan.

Since the second half of the twentieth century the main focus of urbanisation has been in less economically developed countries (LEDCs) and certainly, the immediate future of the world's population will be dominated by a continuation of this rapid urbanisation. Although in many more economically developed countries (MEDCs) these trends are reversing, it is the sheer weight and size of the urbanisation occurring within LEDCs that is tipping the balance of the global population from the countryside towards towns and cities.

1 Defining the Terms

The word 'urban' comes from the Latin *urbs* meaning town. The word 'city' comes from the Latin word *civitas*, which also means town or city. As most urban settlements in Roman times were comparatively small, there was really little distinction made between a town and a city. In the modern world the two terms are still to some extent interchangeable, with the term town used for smaller urban settlements and the term city used for larger ones. In many languages there is no distinction made between town and city, however in the English-speaking world, the distinction varies from country to country.

Historically in Britain, a city was an urban settlement that had a cathedral; yet this has changed in the past 150 years and the definition has therefore become more complex. Thus, although Wells in Somerset with a population of just under 10000 has city status because it has a historic cathedral, Guildford in Surrey, which is a comparatively new diocese with a modern cathedral, still retains the status of a town despite having a population of 127000.

With the growth of huge industrial towns in the nineteenth century, the modern technical definition of a city came into being, and that is simply a town that has been granted city status by a royal charter. This was how places such as Manchester, Birmingham and Leeds first gained their city status. The corporations of large towns still vie for the political 'kudos' of city status, which is a matter of great civic pride and generally allows their mayors to adopt the title of 'Lord Mayor'. The most recent towns in the UK to be granted city status were Brighton and Hove, Inverness, and Wolverhampton in 2000 as part of the new millennium celebrations and Preston, Stirling, Newport (Monmouthshire), Newry and Lisburn (both in Northern Ireland) in 2002 to celebrate the Queen's Golden Jubilee.

A clear distinction needs to be made here between the terms **urban growth** and **urbanisation**, as the two are sometimes confused. Urban growth is the outward and physical development of a town or city into the surrounding countryside and the spreading of the built-up area. By contrast urbanisation is the increasing percentage of people living in urban areas as opposed to rural areas within a country or region. The first example of rapid urbanisation in the modern world was in Britain during the Industrial Revolution, but today it is in LEDCs that the most rapid urbanisation is taking place. Within most MEDCs and even some LEDCs there are now counter currents to urbanisation. **Counterurbanisation** is the process whereby people are leaving cities, particularly inner cities in order to find a better lifestyle in the outer suburbs or in the countryside beyond. As a result of counterurbanisation, a large proportion of MEDC cities are at present losing population. However, in certain major cities such as London, this trend has been reversed. Since the mid-1980s London has been gaining population, particularly as a result of huge redevelopements such as those in the Docklands; this process is known as **re-urbanisation**. Closely associated with counterurbanisation is the process of **suburbanisation**, which has two manifestations: it is both the outward sprawl of suburbs into the countryside and the social, economic and demographic changes resulting from it, particularly on traditional agricultural villages. Although today counterurbanisation is mainly found in cities in MEDCs, in the near future it will become increasingly a phenomenon affecting LEDC cities as their new middle classes choose to live beyond the city limits and to commute longer distances.

2 From Mesopotamia to Megalopolis: A Brief History of Urbanism

a) The Urban Revolutions

Until recently, urban geography texts tended to refer to the 'Urban Revolution' as if it had been just one sequence of events that happened in the ancient Near East (Mesopotamia and Egypt) around the fifth millennium BCE and then spread elsewhere at later dates. Today, the conventional wisdom among urban geographers is that the Urban Revolution was not a single process but took place in three phases: two in the ancient world and one in the modern industrial age.

i) The first Urban Revolution
In the ancient world, the Urban Revolution was not a sudden 'big-bang' type of process, but had its origins in the gradual evolution and change of human economies. For over 3 million years since its evolution, the human population consisted of small roaming communities of hunter–gatherers. Then around 40 000 years ago the gradual discovery of both pastoral and arable farming techniques,

linked partially to the climate changes that took place towards the end of the Ice Ages, led to the **sedentarisation** of populations in places where conditions were most favourable, such as in the river valleys of North Africa and the Near East. The first Urban Revolution took place around 10 000 years ago. These settlements were not the well-ordered city-states that appeared later, but can be regarded as **proto-urban** as they were more advanced than simple rural agricultural communities. Jericho in the Jordan Valley and Çatal Hüyük on the Anatolian plateau, in what is today Turkey, are two of the best-excavated proto-urban sites. These places were fortunate in being close to the sources of some of the most important wild animals that became domesticated (sheep, cattle, goats and pigs) as well as major food crops such as wheat, barley, lentils and nuts.

ii) The second Urban Revolution

The second Urban Revolution was that experienced by the great ancient civilisations of the river valleys of the Near East and South Asia. The areas first affected were Mesopotamia (the land between the Tigris and Euphrates rivers – modern-day Iraq), the Nile Valley (in modern Egypt and the Sudan) and the Indus Valley (in present-day Pakistan). What took place in these areas then spread through the Mediterranean to influence the classical civilisations of Greece and Rome and beyond in later centuries.

A combination of factors changed the proto-urban sites into true urban ones, and these included:

- continued population growth stimulated by large food supplies
- the development of irrigation schemes which would produce larger food supplies
- the merging of existing villages to form larger settlements; this was termed **synekism** by the urban geographer Edward Soja
- the need to build defensive walls against outside invaders.

iii) The third Urban Revolution

The third phase of the Urban Revolution started in Britain at the time of the Industrial Revolution and heralded the beginning of the Modern Age, with its technology-intensive and labour-intensive economic systems. In this period, from around 1710 until 1850, industrial production moved from its water-power-based cottage industry locations dispersed throughout the countryside, to the coal-powered factories located in the rapidly expanding urban areas. It saw the shift in population from a predominantly rural one to a mainly urban one, brought about by heavy rural–urban migration as well as high birth rates within cities. This Urban Revolution started in Britain and then its effects diffused into France, Germany and throughout Europe as well as being transferred across the Atlantic and eventually to all corners of the world. By contrast, in many LEDCs the industrial city is a phenomenon dating back only a few decades. The nature of the industrial or 'modern' city will be dealt with both below and in other parts of the book.

b) The spread of urbanism from the ancient world to the industrial world

From the three so-called 'urban hearths' of Mesopotamia, the Nile Valley and the Indus Valley, where cities had thrived 5000 years ago, urbanism gradually and slowly diffused westwards. By 800 BCE the great cities of Greece such as Athens and Sparta dominated that part of the world; by 750 BCE Greek colonies such as Syracuse were established in southern Italy and they also founded cities around the Black Sea. By around 250 BCE the Greek power had declined and Rome was beginning to dominate the Mediterranean. During the period of the Roman Empire (27 BCE–410 CE) cities were founded all over Europe, North Africa and the Middle East, the majority of them survive to this day. Effectively, England had its first recognisable towns around 150 BCE (fortified Iron Age settlements) before the Romans arrived.

After the fall of the Roman Empire, many European cities continued on a smaller scale, but around 1000 CE there was a great urban revival based on commerce and trade. This gained momentum during the Renaissance period as towns grew and trade developed further with the voyages of discovery and the opening of new trade routes. At the same time, the Reformation in Europe led the way for the development of modern capitalism and the events leading up to the Industrial Revolution. Industrialisation brought with it unparalleled urbanisation and the development of the modern city in Britain, elsewhere in Europe, in North America and subsequently the rest of the world.

Outside Asia and Europe, other civilisations had grown up autonomously and then been repressed and conquered by Europeans. The two most remarkable city-building cultures of the New World were those of the Aztecs based in Tenochtitlán (present-day Mexico City) and of the Incas, based in Cusco in present-day Peru. Both were suppressed by the Spanish *conquistadors* around the turn of the sixteenth century.

c) Cities of the twentieth and twenty-first centuries

In the twentieth and twenty-first centuries, not only have cities become increasingly complex, but so too has the study of urban geography and the multiplicity of terms that it has spawned. The sheer size of cities and the growing number of their functions brought into common usage the terms **metropolis** and **conurbation** to signify very large cities with over a million population. Whereas metropolis (from the Greek 'mother city') is generally applied to large urban areas based on one major centre, such as London or Paris, the term conurbation is more properly applied to a **polycentric** urban area based on two or more centres, such as the Leeds–Bradford conurbation or the Ruhr conurbation in Germany. Whereas the former with a population of 1.4 million is based just on two main centres, the Ruhr, with a population of 3.85 million is based on 15 different urban centres.

Cities with populations of over a million may be referred to as either **million cities** or **millionaire cities**; both terms are equally valid. As the number of such cities increased, a new layer was added to the urban glossary: **megacities**. At present, this term refers to cities with over five million people, but in the future the parameters could be changed; some sources already use the term for cities with over 10 million people.

In the 1970s, the geographer Jean Gottman coined the term **megalopolis**. This concept was based on the huge, almost continuous built-up areas that he recognised developing in the USA and elsewhere. Three such megalopolitan areas were widely written about: 'Bowash', stretching from Boston to Washington via New York, Philadelphia and Baltimore; 'Chipitts', stretching from Chicago to Pittsburg; and 'San-San', stretching from San Diego via Los Angeles to San Francisco. The concept is rather controversial in the US context as there still remains a great deal of farmland and forest between these cities. The megalopolis is most developed in Japan with the near merging of the Kinki (Osaka-based) and Kansai (Tokyo-based) conurbations, especially along the axis of the *shinkansen* (bullet train) and expressway. Perhaps it is therefore more appropriate to use the Japanese term, **obijotoshi**, coined by Ito Ngashima rather than Gottman's megalopolis. In Britain, the megalopolis concept has been the inspiration for one of the most futuristic views of what could evolve in the next few decades. The architect Will Alsop has put forward designs for a SuperCity that would stretch 125 km from coast to coast across northern England; it would follow the M62 corridor from Liverpool to Hull and would include features conceived as large sculptural forms such as the 'Stack' – a fanciful high-rise housing block.

In many countries, the capital city (or the largest city, if it is not the capital), may over-dominate the nation both economically and culturally, because of its sheer size. In the most centralised states of Europe, such as France and Britain, the capital cities have for centuries had very large populations that have enabled them to dominate their territories, not just politically, but also commercially, socially and culturally. In a very large number of LEDCs today, capital cities or main cities are equally powerful. Such cities, when they have populations in excess of double that of the second largest centre of population, are known as **primate cities**. Figure 1.1 shows the occurrence of primacy within Europe and South America. As well as Paris and London, other European capitals that are primate cities include Lisbon, Warsaw and Athens.

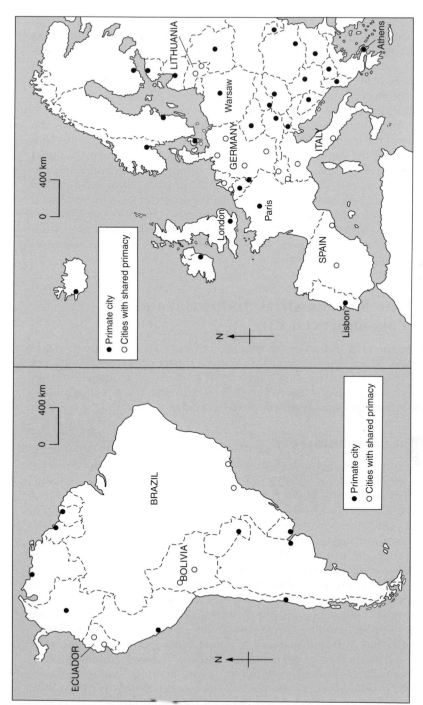

Figure 1.1 Urban primacy in South America and Europe

Within LEDCs, primacy can be regarded as a greater problem, with the snowballing effect of rural–urban migration making main cities ever larger. Frequently, a disproportionate amount of national investment goes into these primate cities on grand infrastructural schemes such as new airports, metros and motorways, which, although they may be necessary, is to the detriment of the funding of projects in other cities.

In some countries, the national wealth is better distributed by the virtue of having **shared primacy**. This is where two of more large cities of roughly the same size dominate the national economy rather than one. Within Europe, Rome and Milan in Italy, Geneva and Zurich in Switzerland, and Vilnius and Kaunus in Lithuania offer examples of primacy shared between two cities. In the case of Germany, a federal country with strong regionalism, the primacy is shared among Berlin, Frankfurt, Munich and Hamburg. As is generally the case within LEDCs, shared primacy is rarer in South America, but is found in Brazil, Ecuador and Bolivia.

3 The Pre-industrial, Industrial and Post-industrial City

Given the long history of the evolution of the city and the many stages of urban development, as well as cultural differences from one part of the world to another, one of the most convenient and academically acceptable ways of classifying cities is into the three categories of **pre-industrial**, **industrial** and **post-industrial**.

a) The pre-industrial city

The concept of the pre-industrial city dates from the work of the US geographer Gideon Sjoberg in the 1960s. His main area of study was in the Islamic cities of the Near East, but the idea of the pre-industrial city today covers European cities before the Industrial Revolution (but particularly in the Mediaeval period), and any city in an LEDC that has not significantly industrialised, e.g. one in which craft and workshop industries still provide the main manufacturing base. Large, obstensively modern, industrial cities may also still have parts to them that are pre-industrial in character. This is particularly well illustrated in many north African cities where there is are extensive 'European' quarters, but at the core there is a *medina* – an old town with narrow, winding streets enclosed by defensive walls. Tunis, Sousse, Sfax and Kairouan, all in Tunisia, are examples. In terms of layout, the most extreme examples of pre-industrial cities are found in West Africa where the dominant element in the city plan is the walled extended family compound; the streets are totally irregular as they are merely the spaces left between the compounds.

The pre-industrial city of Mopti, Mali, West Africa

b) The industrial city

Many of the larger cities of the world, especially in North America and Europe, belong to this category, although they may be at various stages in their transformation towards the post-industrial. Within LEDCs, many of the largest cities also fall into this category, but may have relics of their pre-industrial past positioned cheek-by-jowl with the modern.

Industrial or 'modern' cities are those associated with the rapid urbanisation that takes place as a country is transformed from a pre-dominantly agricultural society to an industrialised one. Segregation of functions becomes much more noticeable. The city centre becomes depopulated and mainly given over to commercial functions. Certain parts of the city develop into industrial zones and the wide range of housing types reflects the big social and economic divisions within the population. With this rapid urbanisation comes the full range of problems associated with contemporary cities: pollution, congestion, waste disposal, unsatisfactory housing, unemployment and crime.

c) The post-industrial city

The post-industrial city is a phenomenon found at the moment almost entirely in MEDCs. These cities have their origins dating back to the second half of the twentieth century as a result of the economic restructuring that took place in many industrialised countries. In Britain, the demise of the large, predominantly industrial city came about as a result of a series of factors in the 1950s and 1960s that led to dramatic changes in the 1980s and 1990s. Many large, industrial areas of cities, as well as their infrastructure, never entirely recovered from Second World War bomb damage. On top of this, many industrial plants in Britain were either old-fashioned and requiring heavy investment or already obsolete. With competition from other European countries, the USA, Japan and then the newly industrialised country (NIC) 'tiger economies' of the Far East, British industry fell behind in many manufacturing sectors that it had dominated in the nineteenth and early twentieth centuries. Cities needed to adjust to other forms of employment in light manufacturing, and above all, in the service sector.

Other MEDCs have followed a similar pattern. In the USA, restructuring was most dramatic in the so-called **Rust Belt** cities of the north and east (e.g. Detroit, Cleveland and Baltimore). From the 1980s

St Mary Axe, London, showing pre-industrial, industrial and post-industrial buildings

onwards, huge shifts were made in urban economies away from traditional, heavy industry towards the service sector. This, as elsewhere, involved the demolition and clearance of large areas of heavy industrial plant and the adjustment of old facilities to new activities. Throughout continental Europe similar changes have been necessary in places as diverse as the Ruhr coalfield cities of Germany and the old industrial areas of Milan.

Nowhere in Britain has the change been greater than in London, and the Docklands redevelopment stands out as its most potent symbol.

Savitch in his 1988 commentary *Politics and Planning in New York, London and Paris* puts it well:

> Post-industrialism can be seen as a transformation of the built environment: factories are dismantled, wharves and warehouses are abandoned, and working class neighbourhoods disappear. Sometimes there is the replacement of one physical form with another – the growth of office towers and luxury high-rises or refurbishing of old waterfronts. Cafés and boutiques arise to feed and clothe the new classes.

The final part of this chapter deals with the post-industrial city in more detail.

4 The Functions of Towns and Cities

Most contemporary towns and cities have a wide range of functions, although their early origins may have been due to one specialised function. Once an urban settlement has been established, it may even lose its initial function or *raison-d'être* (e.g. a mining settlement in which mineral deposits become exhausted), yet it is more likely to continue than to become abandoned; this phenomenon is known as **geographical inertia**. Cities often change their functions and adapt to new economic and political realities, as can be seen in cases of many old industrial cities as they adapt to the post-industrial age. Cities are also influenced by the **multiplier effect** whereby their sheer population size and employment pool enables them to attract new businesses and new functions as time passes. This is why many of the larger urban agglomerations of the world are **multi-functional cities**.

Some of the main original and acquired functions of European cities can be classified into the following 12 categories (for each category one location in Britain and one from elsewhere in Europe will be cited as examples):

(a) Market towns. The vast majority of smaller towns and cities have this as their origin, dating back to periods of population growth and the need to create market centres for the sale of agricultural goods within rural areas; this was particularly the case from the tenth century onwards. Examples include Market Harborough (Leicestershire) and Argenton-sur-Creuse (France).

(b) Defensive sites. In times of civil or international conflicts, many towns were established in defensive sites and fortified with walls and castles. Examples include Caernarvon and Aigues Mortes (France).

(c) Religious sites. Many towns and cities developed around monasteries and cathedrals. Some of these were, and still are, important places of pilgrimage. Examples include Canterbury and Santiago de Compostela (Spain).

(d) Administrative centres. Certain cities evolved as national or regional capitals and therefore some of their main functions are those associated with seat of government. Examples include London and Bern (Switzerland).

(e) Centres of learning. Some cities are dominated by their universities and one of their main functions is teaching and academic research. Examples include Cambridge and Leuven (Belgium).

(f) Route centres. Places where routeways converged have always been in strong positions, both politically and economically. The coming of the railways made some towns and cities more important than they were before. Examples include Crewe and Lyons (France).

(g) Ports. Some coastal and river locations were physically more suitable for the development of sheltered harbours than others and therefore developed into significant ports. These can be put into four subcategories:
- fishing ports, e.g. Grimsby and Mazzara del Vallo (Sicily, Italy)
- commercial ports, e.g. Liverpool and Rotterdam (Netherlands)
- ferry ports, e.g. Dover and Zeebrugge (Belgium)
- naval ports, e.g. Portsmouth and Toulon (France).

(h) Mining centres. Towns have developed on a wide variety of mineral resources, particularly in the period since the Industrial Revolution. Examples include Rhondda (South Wales) and Kiruna (Sweden).

(i) Industrial centres. Most towns and cities are to some degree industrial, but certain urban settlements owe their origins as major centres of population to the changes brought about by the Industrial Revolution. Examples include Manchester and Essen (Germany).

(j) Resorts. With the expansion of the leisure industry in the past century, an increasing number of urban areas are tourist resorts. These can be put into three major subcategories:
- spa towns, e.g. Bath and Spa (Belgium)
- seaside resorts, e.g. Brighton and Benidorm (Spain)
- winter sports resorts, e.g. Aviemore (Scotland) and Zermatt (Switzerland).

(k) Dormitory towns. With the expansion of cities, counterurbanisation and the extension of commuting distances, an increasing number of urban settlements on the fringes of conurbations are becoming dormitory towns, which may have a pleasant environment, but often lack the range of shops and services found in other towns. Examples include Dorking (within the London commuter belt) and Olgiata (Rome, Italy)

(l) New towns. At various stages in history new towns were established, but during the twentieth century there was unprecedented activity in new town foundation as a response to both demographic and economic growth. Examples include Milton Keynes and Latina (Italy).

As can be appreciated from this classification, cities have a greater range of functions than are found in most rural settlements. When examining the actual distributional patterns within cities, land use can be put into six broad categories:

- **commercial** (concentrated mainly in the city centre)
- **industrial** (frequently found in sectors of the city)
- **residential** (normally covers up to 50% of the urban area)
- **recreational** (vary from urban parks in city centres to forests and other wild areas on the city fringes)
- **transport** (includes roads, railways, stations, ports and airports)
- **public utilities** (include water and electricity supplies and, with the exceptions of reservoirs and power stations, may take up a very small area of the total urban surface, as pipelines and cables are generally underground).

Figure 1.2 shows a theoretical arrangement of how these different categories of land use might appear within a typical British city.

5 Functional Segregation in Urban Areas

How and why land use becomes so segregated in cities is due to a variety of interrelated factors, some of which date back to the pre-industrial age, and others have their origins in the modern era. Kostof recognises four main categories of functional segregation and separation in pre-industrial cities, some of which became even more important in the modern world:

- the **administrative district**, where the ruling family resides; this could be seen in Mediaeval Westminster in London, the Louvre area in Paris before the French Revolution, the Forbidden City of Beijing before the end of the Chinese Empire and the colonial city of New Delhi created by the British to administer India
- the **religious district**, which in Mediaeval Europe would have included the cathedral close, such as that still so well preserved in English cities such as Durham and Wells today, or a monastic compound; in other parts of the world, the religious district may be close to a temple or a mosque

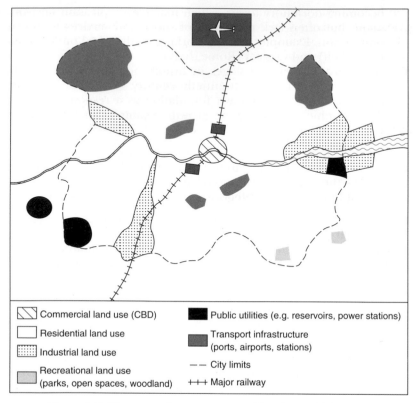

Commercial land use (CBD)

Residential land use

Industrial land use

Recreational land use
(parks, open spaces, woodland)

Public utilities (e.g. reservoirs, power stations)

Transport infrastructure
(ports, airports, stations)

— — City limits

+++ Major railway

Figure 1.2 The general arrangement of land use within a city

- the **commercial district**, which would have been centred the marketplace and the main streets adjacent to it, as well as around port functions of riversides or coastal harbours; the Mediaeval City of London had its main commercial district stretching down from Cheapside and Leadenhall to the various wharves on the Thames
- **residential districts** filled the rest of the built-up area of cities, and there was often considerable segregation within them, based on class, occupation and ethnicity. Where possible, the wealthy would try to locate in the best positions, e.g. in London the Thames-side location to the west of the City was the most prestigious location in Mediaeval times. By contrast the poor had to live in the more polluted industrial areas. Ethnic segregation took place where there were people of different languages, religions and cultural traditions.

There would have been many reasons why these different sectors of cities evolved, but overall it was to do with power, money and influence. Each section of the city evolved primarily to preserve the wealth and influence of each interest group, such as the ruling families, the priesthood and the merchants. However, not all functions were mutually exclusive in pre-industrial cities; for example, the commercial and residential areas overlapped in districts where merchants lived literally 'above the shop'.

With the coming of the modern industrial city, patterns of land use became more complex and diverse. As a result of industrialisation, urbanisation and population growth, many new functions were attracted to cities, as was a more socially and ethnically diverse range of people. Some of the main knock-on effects of these changes were the following factors, which had a great impact on patterns of urban land use:

- New industrial zones of different types grew up in different parts of cities according to their locational needs.
- A much greater polarisation between the rich and the poor within cities created a much greater range of housing types and residential zones.
- Suburbs and urban sprawl extended over much greater areas than ever before.
- The importance of radial routeways (road and rail) and consequent commuting patterns led to radial growth patterns.
- Greater immigration from greater distances led to more ethnic variations in the population.
- Increased commercial activity in the city centre led to the gradual drift of residential populations away from the central business districts (CBDs).

6 Using Land-use Models to Explain Urban Form

From the early twentieth century onwards, theoretical **land-use models** have been widely used in an attempt to explain and understand the complexities of the internal structures of cities. Although such models are often criticised as being too simplistic, their intention has always been as rough guides rather than rigid patterns. For the purpose of examining the different types of land-use models, they can be conveniently classified into three main categories:

- North American urban models
- European urban models
- urban models for LEDC cities.

a) North American urban models (see Figure I.3)

i) The Burgess Model
This was developed by EW Burgess in 1924 and was based on land-use patterns in Chicago. The model is dominated by concentric circles,

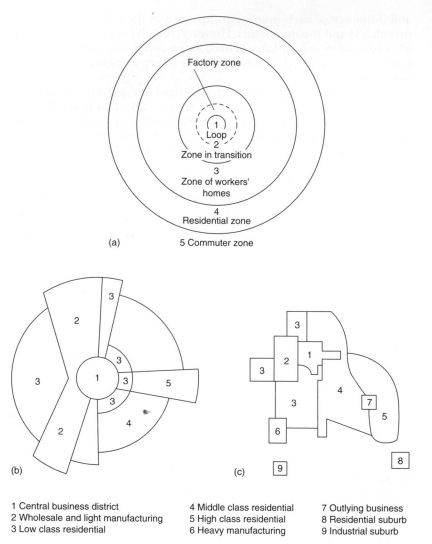

Figure 1.3 The three classic North American land-use models: (a) Burgess'
1925 concentric rings model, (b) Hoyt's 1939 sectoral model, (c) Harris
and Ulman's 1945 multiple nuclei model

which reflect the outward growth of cities. At the same time, land
values decline away from the centre. At the centre is the commercial
core with the highest land values, next comes the zone in transition
with its factories and poor housing, and beyond are increasingly
wealthy and lower density residential zones. As with all models, this is
an oversimplification, but it is of value in explaining patterns.

ii) The Hoyt Model

In 1939, Homer Hoyt put forward his sectoral model based on some 40 different US cities. Hoyt suggested that residential segregation was the main determinant of land-use patterns and, in particular, the rich and poor sections of cities were polarised. Beyond the CBD were sectors of upper and middle class housing located away from the industrial zone, but the lower class housing was, of necessity, close to it.

iii) The Harris and Ullman Model

By 1945, many US cities were experiencing the effects of greater mobility through the motor car and cities were becoming much more decentralised. Harris and Ullman's Multiple Nuclei Model of that date is in many ways a 'non-model' as it is based on a much more fragmented city, rather than one with a rigid pattern. Commercial, industrial and residential functions have examples of 'out-of-city' districts.

b) European urban models (see Figure 1.4)

With a much longer history than US cities, European cities have different structures. In all countries geographers have developed their own variations on the three 'classic' North American models.

i) Mann's Model for a British city

British cities with their long history and more complex social structure need a more detailed explanation than the US models can provide. In 1965, P Mann put forward his model that incorporated both the concentric element of Burgess and the sectors of Hoyt. The basic principles of both are present, but they enabled Mann to represent a much more complex mosaic of housing types which result from social divisions interacting through time.

ii) Model for a Mediterranean city

Northern and southern Europe are very different in their history, their pattern of industrialisation and their social divisions. This model of a large Mediterranean port city shows how the CBD and historic core may be bigger and more residential than in British cities, how the social mix is greater in the more central residential areas, and how the industrial zones (because of later industrialisation) are more restricted. In the case of Mediterranean cities, new upper class residential suburbs are developing well away from the centre in attractive locations, much as they do in some LEDC cities.

c) Urban models for LEDC cities (see Figure 1.5)

It is very difficult to make generalisations about the layouts of LEDC cities, given the great diversity of countries that are represented; therefore models have been developed to explain the urban structure

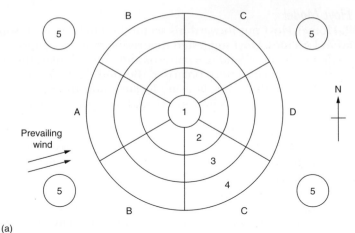

(a)

CBD
1 Transitional zone
2 Zone of small terrace houses in sectors C and D
3 Large old houses in sector A
4 Post-1918 residential areas, with post-1945 development mainly on the periphery
5 Commuting distance 'dormitory' town
A Middle class sector
B Lower middle class sector
C Working class sector (and main council estates)
D Industry and lowest working-class sector

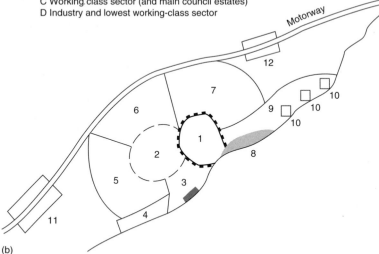

(b)

1 Old historic walled city (mixed commercial and residential)	7 Middle class residential
2 Old commercial district	8 Resort area
3 Port and heavy industry	9 Modern extension of middle class residential
4 Recent port extension	10 High class residential (some gated communities)
5 Low class residential	11 New industrial zone
6 Lower middle class residential	12 New out of town commercial zone

Figure 1.4 Two European city land-use models: (a) Mann's 1965 model for a British city, (b) model for a Mediterranean city

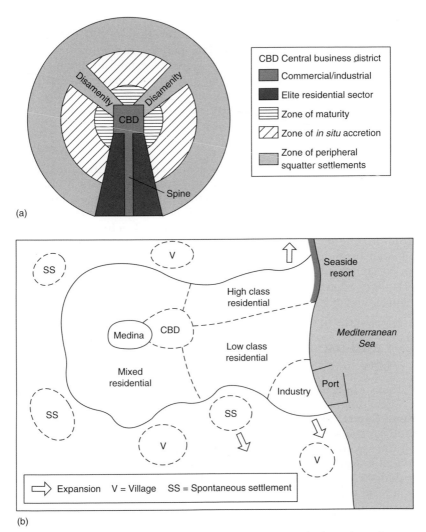

Figure 1.5 LEDC urban land-use models: (a) a Latin American city, (b) a North African city

of typical cities in each of the main cultural zones, three of which will be examined here. Two general comments can be made, however:

- in LEDC cities the wealthier residential areas tend to be further in and the poorer ones further out (the reverse of MEDC cities)
- where the LEDC city has been part of its nation's colonial past it is likely to have a twin CBD or core made up of an original indigenous settlement and a later colonial one.

i) Model for a Latin American city

This model clearly shows the residential 'reversal' of MEDCs with the old élite area close to the CBD. The poorest residential areas are found in both the *periferia* with its low-quality, established homes and beyond, the squatter areas with their unofficial, makeshift housing. The élite groups are, however, through time abandoning the polarised centre in favour of more suburban, safer and often gated community housing.

ii) Model for a North African city

This model shows some of the complexities of north Africa that have both the historic Arab *medina* and the French or Spanish colonial core alongside it; together they form the CBD. Beyond these are areas of mixed housing, with some degree of polarisation between poorer housing nearer the port and industrial zone and richer housing away from it. Higher class housing is also found in seaside suburbs. Squatter housing or *bidonvilles* have largely been cleared to make way for social housing schemes. Some low-quality housing remains on the periphery where villages are being absorbed into the city's fabric. Tunisian cities such as Sousse and Kairouan fit this model.

7 Land-use Models and the Post-modern City

Perhaps the ultimate land-use model is that of Dear and Flusty, two US geographers working in Los Angeles. In 1998 they produced the **Keno Capitalism** Model that represents the urban structure of post-modern cities (see Figure 1.6). Not surprisingly, the model is based on the way in which their home city is evolving. The base map of the model is a grid pattern (which reflects the criss-crossing of freeways within the urban fabric of Los Angeles) and the word 'keno' relates to a form of lottery. What Dear and Fluty are suggesting is that capitalism will influence what locates where within the city in a totally random way. Unlike previous land-use models, there is no real logic behind what locates where, and the processes that cause related functions to cluster into the same place or incompatible functions to be polarised on opposite sides of the city are absent. It is also a model that shows total urban decentralisation.

The Dear and Flusty model has been highly criticised as being based so much on Los Angeles and that whereas some other US cities, particularly in the western states, may to some degree be evolving in the same way, most are not. The model is even less applicable to countries where cities have a long history and are still dominated by the effects of centralisation, as is the case within most of Europe.

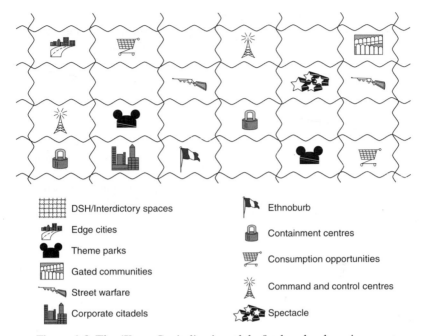

DSH/Interdictory spaces

Edge cities

Theme parks

Gated communities

Street warfare

Corporate citadels

Ethnoburb

Containment centres

Consumption opportunities

Command and control centres

Spectacle

Figure 1.6 The 'Keno Capitalism' model of urban land-use in a post-modern city. After Dear and Flusty (1998).

CASE STUDY: LOS ANGELES AND POST-METROPOLITAN URBAN FORM

A large amount of academic study has been carried out in recent years on Los Angeles. Many urban geographers regard it as the most developed city in the world in terms of its structure (or lack of it), its relationship with the motor car and the problems it encounters. The land-use model that most clearly helps to explain the urban patterns of Los Angeles is the 'Keno Capitalism' model discussed earlier.

Los Angeles may be regarded as a key to the pattern of urban life that may be expected elsewhere in the future, but there are many obstacles to the Los Angeles model being followed in parts of the world beyond North America. In Western Europe, for example, well-established cities with centres rich in architectural heritage are less likely to follow the model, but despite this there are signs that newer wealthy suburbs of cities in countries such as Spain and Italy are beginning to resemble their Californian counterparts.

Edward Soja's book entitled *Postmetropolis* is the most detailed recent study of Los Angeles. The very concept of a post-metropolis is one of a city that has undergone the transition from an industrial phase to a post-industrial one, and has then gone one stage further. As was discussed above, Los Angeles is a formless city that fits no conventional models and its land-use pattern has evolved by chance as much as from any comprehensive planning. Moreover, in its amorphousness it is no longer recognisable as one city.

Soja recognises six main phenomena that are each partially responsible for the post-metropolitan transition. These are:

- the globalisation of capital, labour, culture and information flows
- the *Post-Fordist* economic restructuring that has involved a transition from old heavy industries to high-tech and tertiary activities
- the restructuring of urban form; this involves the continuing decentralisation of cities as their suburbs grow, and the development of new forms of outer cities such as **edge cities**
- social restructuring within cities, with the urban mosaic becoming even more fragmented than in the past, with gentrification on the one hand and greater poverty, homelessness and the emergence of an underclass on the other
- the emergence of **carceral cities** as a consequence of greater social polarisation; these are the gated communities of the rich that largely guard and exclude the wealthier classes from the outside 'real' world, and therefore from crime, poverty and generally having contact with the more marginalised elements within society. As Soja puts it, '[these communities are] where police replaces *polis*'
- the blurring of the edges between real places and imaginary places. Real cities with their post-modern architecture are coming to resemble the computerised urban landscapes of the **Simcity** (the popular computer urban land-use simulation game).

Questions

1. With reference to specific examples, explain the differences between the terms **conurbation**, **metropolis** and **megalopolis**.
2. Outline the main differences between **industrial** and **post-industrial** cities.
3. Why is there such competition for space between different functions within contemporary cities?
4. Assess the value of the use of **land-use models** in explaining the layout of cities in i) North America, ii) Europe and iii) LEDCs.

2 Old Towns, New Towns and the Role of Urban Planning

KEY WORDS

Garden City the type of idealised low-density town envisaged by Ebenezer Howard at the end of the nineteenth century
New Urbanism a late twentieth-century planning movement in the USA
urban palimpsest the superimposition of different street patterns belonging to different historic periods
utopian tradition the planning tradition of which Garden Cities were a part

'The planners have grasped a single truth. They have recognised that in the city they are dealing with some hugely enlarged frame for human behaviour in which moral extremes are likely to be the norm. The city, they sense, is the province of rogues and angels, and a style of building or a traffic scheme, might tip it conclusively in one or the other direction.'

Jonathan Raban *Soft City*

Urban planning has been around almost as long as cities themselves. The shape and internal form of a city may be the product of an accident or design; frequently it is a compromise between the two. Since the Industrial Revolution, however, there has been a much greater divergence of views, and we now live in a world in which, as suggested in the quotation from Raban, planners and their schemes are judged by the general public as being either good or bad. Moreover, the effect their plans may have on public behaviour is very great indeed.

1 The Concept of the Ideal City

The concept of the ideal city has appeared in literature for at least 1000 years. In Western Europe, the inspiration often came from Biblical descriptions, in particular those of Solomon's Temple in Jerusalem and the vision of Heaven in the Book of Revelations. Books on ideal cities and societies within Western literature include St Augustine's *City of God* and Thomas More's *Utopia*. In poetry, many visionaries such as John Milton and William Blake also touched on the theme. Renaissance artists had visions of ideal cities, and perhaps the most famous of these is the image of an ideal square in a city by the Italian artist Piero della Francesca. At the same time many architects were creating plans of their ideal cities.

2 The History of Urban Planning

The remains of the earliest cities in the ancient Near East, such as Ur and Babylon (both in present-day Iraq) show few signs of overall systematic planning. The grouping of important buildings and certain ceremonial streets do show some planning.

The first known town planner was Hippodamus of Miletus, in Greece, who was operating several millennia later, during the first century BCE. Throughout the Mediterranean world there are grid-pattern cities laid out according to the theories and practice of Hippodamus. These include the ancient sites of Selinunte in Sicily, Pergamum in Asia Minor (present-day Turkey) and Jerash in Jordan.

The Romans took town planning one stage further and their cities used a rigid form of the grid pattern. The works of the architect and town planner Vitruvius, the ten books entitled *De Architectura* ('About Architecture'), provide a great amount of detail about how the Romans planned and laid out their cities. The foundation of the cities was surrounded by religious and superstitious rites, whereas the laying out of the internal structure was very logical and formal. Once the line of the city walls was established, the surveyors moved in and using an instrument called a *groma*, laid out the streets in a rigid grid pattern. Each block or *insula* was almost exactly the same size. The city was dominated by two main streets: the east–west *decumenus maximus* and the north–south *cardus maximus*. Close to the intersection of these two vital arteries, the forum was located; this acted as both commercial and administrative centre.

Throughout the regions that had been under Roman administration cities with grid patterns of streets, or the remains of them, still exist. Examples include Aosta, Lucca and Àscoli Piceno in Italy and Chichester, Gloucester and Silchester in England.

The cities of post-Roman Europe were in general typical pre-industrial towns, with a lack of planning and irregular, organic street patterns that included alleyways and dead ends. In some cases this irregularity evolved out of a Roman grid pattern, and this can still be seen by analysis of town plans in such places as Palermo, Sicily, Lucca, Tuscany (both in Italy), as well as in Chichester, Gloucester, Colchester, Canterbury and Leicester in England. As Abercrombie noted:

'... whereas many of our towns have a Roman skeleton, they are clothed with the flesh and blood of the Middle Ages'.

Figure 2.1 shows the relationship between Roman and Mediaeval street alignments in Lucca, Italy.

Such changes in street patterns through the ages has enabled us, through town plan analysis, to 'read' and interpret the way in which urban areas have evolved. Just as in rural areas, a town's landscape can be seen as a sequence of historical–cultural layers imposed on the

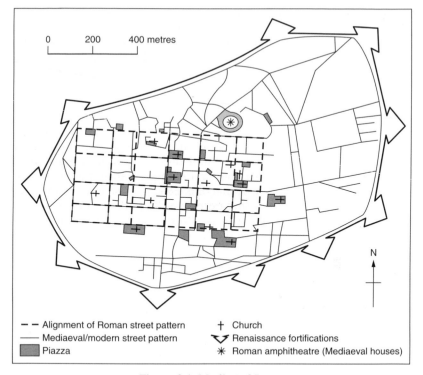

0   200   400 metres

-- Alignment of Roman street pattern   † Church
── Mediaeval/modern street pattern   ∨ Renaissance fortifications
▓ Piazza   ✳ Roman amphitheatre (Mediaeval houses)

N

Figure 2.1 Mediaeval Lucca

original surface through time; this concept, using the analogy of a reused parchment, is known as the **urban palimpsest**.

Alongside the existing cities based on their earlier foundations, the Middle Ages saw the foundation of a large number of new urban settlements in Europe. The most active period of new town development was in the late thirteenth and early fourteenth centuries, and coincided with a time of both rapid population growth and considerable military activity. These new towns were generally well fortified and laid out on a rigid grid pattern in order to make them easy to defend. The general name given to such settlements is the French word *bastide* (from *bâtir*, to build). As Burke puts it:

> 'A golden age it was. In scarcely a lifetime many hundreds of new towns were built and occupied all over Europe, from Vilnius to Bordeaux and Berwick-on-Tweed to Berne. They were built by kings and nobles to defend newly acquired territories and to attract settlers to exploit the land and develop trade; and some were set up in rivalry to established trading points.'

In Britain it was the English King Edward I (1272–1307) who was most active in the foundation of bastide towns. The most impressive of these were the castle towns of North Wales, such as Conway and Caernarvon, built during Edward's campaign to subjugate the Welsh. Edward also saw the building of Berwick-on-Tweed as a defensive town on the borders with Scotland. Some of Edward's new towns were built for peaceful purposes and these included the new ports of Kingston upon Hull and Winchelsea.

The Renaissance, which started in Italy around 1450 but took a long time to diffuse throughout Europe, brought with it new approaches to town planning. Various architects, particularly in Italy, designed 'ideal cities', many of which had two main attributes: they had a perfectly symmetrical layout and they had the new star-shaped fortifications which provided better defence against the improved and more powerful artillery of that age. Few of these cities were actually built, but two examples that do exist are Palmanova near Venice (see Figure 2.2) and Naarden in the Netherlands, both built in the sixteenth century. The ideas of the Renaissance continued through to the beginnings of the nineteenth century. Many European cities were rebuilt or extended using these grand plans and geometric patterns. In Britain of the eighteenth and early nineteenth centuries two of the most important schemes were the development of Bath and Edinburgh New Town with their new geometrical streets, squares, circuses and crescents. Within London the development of Regent's Park, Great Portland Street and Regent's Street in the West End was

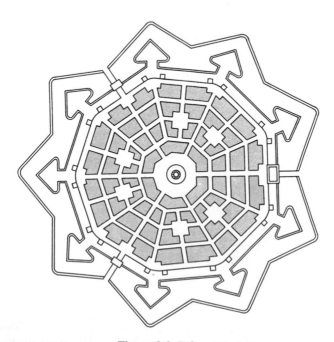

Figure 2.2 Palmanova

the boldest statement of this style of replanning. There were also schemes in Britain that were never built, such as the grand design of Sir Christopher Wren for the redevelopment of London following the Great Fire of 1666.

3 The Main Trends in Modern Urban Planning

The theory and practice of modern town planning have their origins in the excesses of the Industrial Revolution. It was the changes brought about by urbanisation and industrialisation during the eighteenth and nineteenth centuries that produced these excesses of overcrowding, poor housing, disease and pollution. Cities expanded rapidly during this period and in a piecemeal fashion; there was rarely an overall plan to the city itself, but only to the individual sections of cities as they were added to the urban fabric. Thus, nineteenth-century cities tended to be a patchwork of neighbourhoods, each with factories and grid patterned streets of dense terraced housing, with each neighbourhood being developed in a piecemeal way. The general name given to this sort of development in Britain is **by-law housing**, as a local law had to be passed to enable each area of former farmland to be built on. Very typical within these new residential areas was '**back-to-back**' housing. Hoskins comments on the situation in Nottingham where industrial city housing was extremely poor. Population densities there were over 400 per hectare, and the rows of three-storey back-to-back terraced houses backed on a courtyard with an open drain running through it; lavatories were communal with around 12 of them shared by the large families from around 40 houses. Access to the internal courtyard was through narrow tunnels, and the only other facilities each block would have been likely to have were a corner shop and a pub. Not only did this housing create an appalling environment in which to live, but the rents charged for it took up a large proportion of each family's income.

Burtenshaw (1991) recognises six main traditions within European urban planning which date from the nineteenth century or earlier and have had a profound influence on planning in the modern era. Some of these traditions developed as a reaction to the appalling conditions of industrial cities with their congestion from high population densities, high pollution levels from uncontrolled industrial development and substandard, cheaply constructed, speculative housing, much of which quickly degenerated into slums. These traditions were also to have an effect on city planning throughout the world, as they evolved at a time when Europe's cultural influence was at its peak. The six traditions are:

* The **Authoritarian tradition**, which followed the practices of the Renaissance in the laying out of grand geometric plans and street patterns; the grand boulevards of Hausmann's late nineteenth-

century Paris and the schemes of the Fascist era in Italy are both examples.

- The **Utilitarian tradition**, which was a *laissez-faire* approach to urban change, going along with rapid urban growth and often producing poor-quality large-scale housing, as in many of the mining towns of South Wales and in city estates such as Somers Town in Camden, London.
- The **Utopian tradition**, which followed the quest to produce an ideal society and was to be a very significant tradition in Britain. From the late eighteenth until the late nineteenth century, various philanthropic industrialists who were often socialists or Quakers established well-planned towns or suburbs with factories, decent housing and a wide range of facilities for their workers. Examples of these ideal societies include New Lanark near Glasgow founded by Robert Owen in 1785, Saltaire near Bradford founded by Titus Salt in 1851 and Bourneville, a suburb of Birmingham founded by the Cadbury family in 1878. In literature, the character of Oswald Millbank in Disraeli's novel *Coningsby*, a philanthropist–industrialist, built in this tradition. In 1898, the ideas of the utopian tradition came together in Ebenezer Howard's book entitled *Tomorrow*. In this he put forward his principles of what the ideal community should be like and in doing so founded the **Garden City** movement. Among Howard's most important principles were:

  - garden cities should be limited in size to 32 000 people
  - they should be self-contained with sufficient jobs
  - there should be plenty of open recreational spaces and a green belt of farmland around them
  - there should be a great range of social institutions (e.g. schools, libraries, sports clubs)
  - there should be segregation of land use, particularly between industry and housing
  - land was to be owned by the municipality and rented out, thereby controlling land values.

  Although few places were actually built following Howard's principles (Letchworth in 1901 and Welwyn Garden City in 1920), the ideas were to some degree adopted in the setting up of British New Towns after the Second World War, dealt with later in this chapter.

- The **Romantic tradition**, which looked back to the great European cities of the past and the arrangement of important buildings and vistas focusing on them. The main proponent of this idea was Camillo Sitte who worked on developments in Vienna and Madrid; in London, the late nineteenth-century Trafalgar Square can be regarded as part of this tradition.
- The **Socialist tradition**, which is based on the interpretation by national and local government of the writings of Marx and Engels

and the action they have taken in creating master plans for cities and in particular creating good-quality public housing. It is generally accepted that these principles were rarely successfully achieved in Eastern Europe during the Communist era, but that cities such as Bologna in Italy, which has had decades of left-wing administration, have come closer to the principles.

* The **Technocratic tradition**, which took the modernist line in urban development. France came closest to adopting these principles with the plans of Garnier and le Corbusier. The radical principles involved large-scale high-rise public housing, bold radial and grid pattern layouts and large industrial zones. Many Parisian suburbs reflect this tradition.

4 New Towns: Past, Present and Future

The concept of the 'New Town' is not new. As mentioned above, planning is almost as old as cities themselves. New Towns belong to every age and their foundation generally coincided with periods of population growth and the colonisation of new lands. However, there are as many reasons for new town building as there are functions of towns and cities. Planned settlements in Europe and beyond are to be found associated with a wide range of activities, including military and strategic functions, the development of new resources such as mineral deposits, the evolution of new route centres with the coming of the railways, the establishment of spas and seaside resorts, the foundation of new ports, the creation of new capital cities and administrative centres, as well as the development of the purely residential function in the case of new commuter-belt towns.

CASE STUDY: FASCIST NEW TOWNS IN ITALY AND ITS FORMER COLONIES

The front cover of this book shows one of many metaphysical cityscapes painted by the Italian artist Giorgio de Chirico. Dating from 1915 and entitled *L'Enigma di una Giornata* (The Enigma of a Day), it has a dream-like quality, with its simple, stark arcading, long afternoon shadows and an almost deserted main town *piazza*. Within three decades from the date of the painting, over 70 New Towns had been established in Italy during Mussolini's Fascist regime. Architecturally, these Fascist towns have an atmosphere similar to that displayed in a de Chirico cityscape, and as one critic noted, the artist:

'anticipated a new world in which myth and the contemporary can coexist, in which time and history have stopped ... making the real seem unreal'.

When Mussolini came to power in 1922, he saw himself as a new Caesar and like the emperors of ancient Rome wanted to make an impact through great public buildings and other works. At the same time, the Fascists were highly influenced by the Futurist Movement in the arts. Futurism celebrated the new technology of the time, rejecting what had gone before. Aircraft, sports cars and the machinery of war were popular themes by Futurist artists, and the streamlined vision of this future is as much part of the architecture of the Fascist New Towns as is the nostalgia for the grandeur of ancient Rome.

The 1920s and 1930s were a period of rapid population growth and restrictions on emigration to the USA and elsewhere; it was also a time during which the use of natural resources was being developed in an unprecedented way. The New Towns provided settlements in the areas where these changes were being made and were established between 1927 and the fall of the regime in 1943. Although these new settlements include functions as diverse as that of Carbonia, a mining centre on the Sardinian coalfield, and Tirrenia, a seaside resort in Tuscany, the vast majority of New Towns were associated with land reclamation. In his 'Battle for the Grain', Mussolini embarked on massive land reclamation schemes in many parts of Italy in order to increase food production and therefore make the country more self-sufficient.

A mosaic showing Mussolini harvesting grain and one of his new towns in the background

The biggest new town and reclamation project was that of the Agro Pontino (the Pontine Marshes) lying along the Tyrrhenian coast to the south of Rome. Here, between 1926 and 1935, 75 000 hectares of land were reclaimed from the marshes to create arable farmland on which a planned settlement hierarchy of new farms, new hamlets and new towns was established. Some 35 000 farms were established for individual families, 24 *borghi* (agricultural hamlets providing for the everyday needs of the local farmers) and, at the top of the hierarchy, were the four new towns which acted (as they still do today) as major service centres for the whole area. The names of the towns are evocative and nationalistic: Latina (after the Latin peoples), Sabaudia (from the name of the then Italian royal family), Aprilia (after the month of the mythical foundation of Rome) and Pontinia, from the name of the region itself.

The layout of these new towns took a similar form to that of those built elsewhere:

- they have radial and/or cellular forms to their overall plans
- some of the main roads are broad, tree-lined, quasi-ceremonial boulevards
- major buildings such the main church, town hall and post office are grouped close to the centre
- the focal point of the street layout is a large, central *piazza*
- there are several large parks and other open spaces close to the centre.

As, in theory, an egalitarian society was being created, housing type did not vary a great deal, and at least when these new towns were first founded, there were not great variations in housing quality; neither were there noticeable rich and poor sectors. Even today, there appears to be little polarisation of social areas.

Latina, the provincial capital and largest of the new towns, now has a population of over 106 000, making it the second largest city in the Lazio region, after Rome itself. Founded in 1932, it still retains its original urban plan (see Figure 2.3) conceived by the architect and town planner, Oriolo Frezzotti. The radial pattern owes much more to the ideal cities of Italy's Renaissance and Baroque past than anything of ancient Rome. The great monumental buildings on the other hand are evocative of the Roman past. The wideness of the central streets has enabled Latina to cope relatively well with modern motor traffic and some of these have effectively formed an inner ring road. The web-like structure of the plan has enabled the city to develop in a concentric and cellular manner. The heavy industry associated with Latina is kept well away from the city as it is near the railway station that is 10 km to the north. Light industry is mainly strung out along the arterial roads on the outskirts of the town,

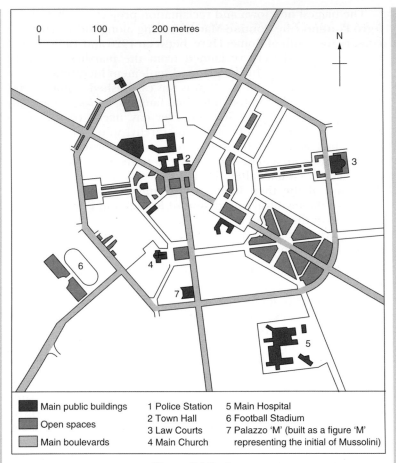

Figure 2.3 Latina

thus creating a great deal of segregation between industrial and other functions.

Not only did Italy build these Fascist new towns within its home territory, but they were also planned and planted in various parts of the former Italian Empire. New towns and cities were built throughout Libya, Eritrea and Ethiopia. Many of these settlements have been altered since they were built, or partially demolished because of what they represented. The Eritrean capital Asmara, however, has so much Italian architecture of the 1920s and 1930s, which is very well preserved or under restoration, that it has become a major selling point in Eritrea's fledgling tourist industry. The Italian-built hotels, cinemas, churches, garages, restaurants and coffee bars give the city a collection of modernist structures unique in the tropics.

CASE STUDY: THE RENEWAL OF BRITISH NEW TOWNS

After the Second World War, 28 New Towns were established to deal with overspill from London and other large cities, to regenerate old industrial areas as well as to encourage development in areas not so densely populated (see Figure 2.4). A total of 1.4 million people were accommodated in these new towns.

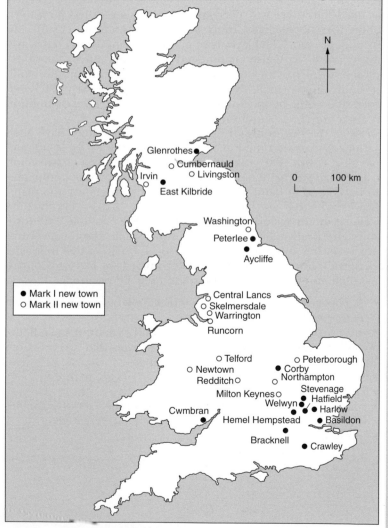

Figure 2.4 The location of Britain's New Towns

Although the New Towns were following in the tradition of Garden Cities and the theories of Ebenezer Howard, many of them fell short of his aspirations. Although the towns were originally to have populations of 20 000–60 000, some such as Basildon have gone well above twice the upper limit. New Towns were also meant to be self-sufficient in employment, but large-scale commuting from them to neighbouring places is common.

Above all, it is the poor quality of much of the building that has caused problems. New Towns were built cheaply and quickly during a period of post-war austerity and therefore now fail to meet the needs of their twenty-first century inhabitants. As early as the 1960s there was a recognised syndrome called 'new town blues' with people living in these uniform communities suffering from depression. The layouts of many of the housing estates and town centres have also encouraged high crime rates. One of the most critical findings about these towns was an opinion poll taken in Cumbernauld New Town in Scotland in early 2005; well over 90% of the people questioned favoured the total demolition of their concrete community.

Since the beginning of the new millennium there have been various initiatives aimed at regenerating some of these 50-year-old New Towns. The government has established an organisation called English Partnerships (EP) that enables private companies to work hand in hand with town councils on regeneration plans. New Towns that are currently undergoing 'makeovers' include Telford in Shropshire, Harlow in Essex, Stevenage in Hertfordshire and Bracknell in Berkshire; in particular, the town centres are receiving attention. In the case of Harlow, the following changes are being made (from 2002 onwards by Harlow Council in conjunction with Wilson Bowden, their EP partner):

- 6.5 hectares of land are being redeveloped to extend the town centre by 35%
- new superstores, retail warehouses and restaurants are being built in more attractive architectural styles that are less drab than the original town centre buildings
- 1200 additional parking spaces
- a new bus terminus development
- new public facilities including 'Shopmobility' services for disabled people and greater use of surveillance cameras
- coverage to combat crime.

CASE STUDY: CELEBRATION, FLORIDA: A DISNEY CORPORATION NEW TOWN

In the late 1980s and early 1990s in the USA, various planners and architects formed what is known as the **New Urbanism** movement. Many of the objectives of this movement were connected with putting right the problems that had been caused in previous decades by the rapid growth of car ownership and therefore the domination of urban areas by motor traffic. Cars had caused old-fashioned neighbourhoods to be replaced by sprawling, anonymous suburbs that lacked community focal points. New Urbanism has aimed to re-introduce walkable, human-scale communities back into cities. In 1981, the prototype New Urbanist town was built at Seaside, Florida. Other examples that followed were Harbor Town, Memphis, Tennessee, and Mashpee Commons in Massachusetts. It could be argued that a British equivalent of the New Urbanism is the small rural town of Poundbury in Dorset, where, in 1991, the Prince of Wales established a new settlement designed by Léon Krier. It uses traditional materials and each house has a different design in order to give the appearance of a place that has undergone historical evolution rather than that of a new housing estate with buildings all in the same style. As with the New Urbanism in the USA, it follows both romantic and utopian planning traditions.

In 1994, the Disney Corporation added the new town of Celebration in Florida to the list of New Urbanist settlements. In keeping with many of the other projects built by the corporation, it mixes fantasy with reality. Celebration 'celebrates' historic small-town America and harks back to an idyllic and innocent past that is firmly embedded in the national psyche, but in reality never existed.

There are eight neighbourhood communities envisaged in the first phase; homes are built in a variety of 'approved' styles, including Victorian, Mediterranean, French, Coastal and Colonial Revival. Homes are also built in seven sizes from small bungalows to mansions. As most people will live within a 2-km radius of the downtown area with its traditional-looking shops, churches, post office, banks and theatre, a lot of people will be walking or using bicycles, Segway scooters or small electric cars rather than the large vehicles associated with most US suburban dwellers. Celebration is therefore expected to become a healthier and less polluted town than elsewhere in the USA.

The first 351 home plots were sold quickly in 1995 and there was soon a long waiting list for those wishing to live in Celebration. The town is far from a gated community and its downtown area relies heavily on visitors for its income; various community events take place in the centre throughout the year including Christmas festivities with artificial snow in this sub-tropical climate. By 2004, Celebration had over 9500 inhabitants living in 3745 households. Far from being a cross-section of US society, the new town is 94% White and with average incomes reflecting a mainly middle class professional population. Also 7% of the population is over retirement age. The median age of inhabitants is 37 years and 45% of households have children under the age of 18 living with them.

Towns such as Celebration, and the New Urbanism movement in general, have had their critics. Some of the main negative comments about them have been that:

- they put aesthetics before practicality
- they ignore many of the best principles of modern urban design and go back to pre-1920s principles
- they are built just out of nostalgia for the past
- the towns have all been built on green sites and therefore have merely been yet another form of urban sprawl.

Questions

1. Outline the main ways in which historic attitudes towards planning have influenced the development of contemporary cities.
2. Using the quotation at the beginning of this chapter, explain how planning decisions can greatly influence the lives of people living in cities.
3. What are the advantages and disadvantages of creating New Towns within countries?
4. Assess the role of the Garden City movement in how cities are planned today.

3 The City Centre and the Central Business District

KEY WORDS

bid rent curve (also known as the **Alonso Model**) a graph which explains the relationship between rental values, proximity to the city centre and land use

core and frame a type of CBD which has an inner and an outer zone, each with different types of land use

devolution the movement of functions away from a central area

Peak Land Value Intersection (PLVI) the point in the city centre where land values are at their highest

polycentric CBD the CBD of a very large metropolitan area that has different subsections, each with a different range of functions

'The axis of the earth sticks out visibly through the centre of each and every town and city,'

Oliver Wendell Holmes *The Autocrat of the Breakfast Table*

Whereas suburbs are generally anonymous in their character, city centres are strikingly obvious. Architecture, building lines, the functions of buildings, the nature of their public spaces, and traffic and pedestrian densities are all telltale signs of city centres. As city centres are by their nature the most accessible places within an urban area, the most important functions that need to be accessible cluster there. This in turn leads to greater competition for space and a consequent rise in land values and property prices. One of the greatest consequences of this has been the development of high-rise buildings by which landowners maximise their profits within limited spaces. The city centre has thus become both the powerhouse and the most potent symbol of a metropolitan area.

1 The Evolution of the Central Business District

In the pre-industrial world, cities were not as dramatically divided into centres and suburbs as they are today, but the concentration of important functions was there for all to see. In the ancient cities of Mesopotamia and Egypt, the main religious buildings dominated cities but were not necessarily accessible to all. Roman cities had their most important functions, the *forum* (market place), *basilica* (administrative building equivalent to a town hall) and main temple, all located close to the intersection of the two main urban thoroughfares. In Mediaeval European cities this pattern merely continued with the town hall or feudal lord's palace as the main administrative, secular building and the cathedral or collegiate church as the most important religious building, both positioned overlooking or close to the market place, effectively the hub of commercial wealth. Within the traditional Islamic cities of north Africa the close proximity of the bazaar, main mosque and fort or *kasbah* is similar to the pattern of the Mediaeval European city.

In the industrialised world of nineteenth century, just as in the Middle Ages, prestigious buildings were erected to show off the wealth of cities. Town halls, corn or other commodity exchanges, banks and insurance offices were built on a palatial scale and in a range of eclectic styles. Such multi-storeyed buildings were present in large concentrations and mere city centres had actually become transformed into true central business districts (CBDs). London, Birmingham, Liverpool, Manchester, Leeds, Melbourne, Montréal and Mumbai are all examples of great cities that still contain within their CBDs many commercial and public buildings reflecting the confidence of the Victorian Age. Beyond the influence of the British Empire, similar grand-scale buildings went up during the industrial age; within continental Europe, cities such as Barcelona, Milan, Helsinki and Riga contain a wealth of nineteenth- and early twentieth-century commercial and civic buildings of an opulent style.

The twentieth century is often referred to as 'the American Century' and it was the USA that had the most dramatic influence on the form of CBDs. The evolution of high-rise buildings in order to make maximum use of limited space within the central areas of New York and other US cities gradually became the norm in the twentieth century. Despite the retreat from high-rise housing in the UK and the events of 11 September 2001 in the USA, in many CBDs of the twenty-first century, in both MEDCs and LEDCs the trend is to build higher and higher. Cities such as Kuala Lumpur, Taipei, Singapore, Hong Kong and Shanghai in Asia are taking the 'American Dream' one stage further as they vie with each other for more futuristic high-rise designs from the world's top architects.

CASE STUDY: THE CHANGING LOCATION OF CENTRAL CAIRO

Each historical phase in urban development has its own locational rationale as well as different planning regimes and building styles; Cairo, with its history stretching over 2500 years, is a remarkable city as its 'centre of gravity' has changed so many times. Figure 3.1 is a schematic map of how Cairo's central area has evolved.

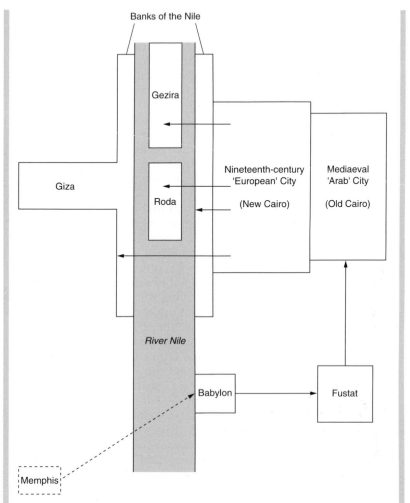

Figure 3.1 The changing location of the centre of Cairo

Memphis was the first important city in the area of today's Greater Cairo. Located 15 km to the south of the current city centre on the left bank of the Nile, it was the capital of the united kingdoms of Upper and Lower Egypt, at the height of its power some 2500 years ago. The centre shifted in Roman times to Babylon-in-Egypt, the old city of Cairo 3 km from the city centre on the right bank. This is now one of the poorer districts of the city where there is a large concentration of Coptic Christians. In the seventh century, after the Muslim conquest, a new city was built to the east of Babylon at Fustat. Today, this has also become degraded into a poor district with a huge number of squatters living close to the smoking furnaces of the city's brick and ceramic kilns. The Arab conquerors then built two new cities further north, *El' Qata'iyeh* in the area around the Ibn Tulun Mosque and *El Qahira*, 'the Victorious', which gave its name to Cairo. These two areas are known generally as the 'Arab City' and include today most of Cairo's historic core, with mosques, bazaars and merchants' houses. Until the late nineteenth century, this was the main commercial and administrative centre of the city and the home of the middle classes as well as the shopkeepers. Today, it has a high population density, the poor have moved in and many merchants' houses are subdivided into squalid living quarters.

In the nineteenth century, the gradual taming of the floodwaters of the Nile enabled the shift of the CBD westwards into the so-called 'European City'. This was built by French and British colonialists on a geometric street pattern, with banks, large stores, luxury apartment blocks, offices, cafés and restaurants as major types of land use. Further taming of the floodwaters in the twentieth century, including the building of the Low and High Dams at Aswan, enabled another phase of development even closer to the banks of the Nile. Today, the biggest concentration of high-rise offices is on the right bank of the Nile, which is now the core of the CBD. Other important areas with CBD functions as well as hotels, restaurants and luxury housing include the two islands of Roda and Gezira, the left bank of the Nile and the road which leads from the left bank to the pyramids at Ghiza. The road to Ghiza became one of the great leisure and entertainment areas of the Middle East when Beirut became a no-go city during the Lebanese civil war, but now Cairo, in turn, is losing out to Dubai as the 'playground' of the region. The high-class residential area of Heliopolis that is located close to the airport also has numerous commercial functions and can be regarded as a detached part of the CBD.

2 Defining and Delimiting the CBD

The edges of CBDs are more clearly marked in North American cities than they are in most European countries, and it is therefore not surprising that much of the earliest study of modern CBDs was carried out in the USA. Murphy and Vance, the North American urban geographers working in the 1950s, were the first academics to carry out detailed studies on CBDs. They were primarily concerned with the defining and delimiting of city centres. British, Scandinavian and

German scholars followed their example and a whole range of techniques for defining CBDs evolved. Using such techniques to survey a central city area, the statistical information could then be plotted on large-scale maps in order to make a clear definition of the CBD itself. These techniques can be put into three broad categories:

- Those based on **rental values** or **land values**; one such formula, developed by the Swedish geographer Olssen, is the **Shop Rent Index (SRI)** which is calculated by dividing the total shop rents of a building by the length of frontage of the building. Similar formulae have been developed for offices and other CBD functions.
- Those analysing **building heights**; the most commonly used example of this is the **Central Business Height Index (CBHi)** developed by Murphy and Vance. This is calculated by the total floor area given over to CBD functions within a building by the total ground floor area.
- Those analysing **building functions**; such techniques are more subjective as they involve categorising buildings into different groupings (e.g. banks and other financial offices, retailing, public buildings, residential) and then plotting the findings on large-scale maps by, for instance, colour coding. From such a survey the CBD can be delimited as being the area with over a certain percentage of commercial functions. The range of categories chosen for a city in one country may not necessarily be appropriate elsewhere.

3 The Structure and Land-use Patterns of the CBD

The importance of centrality, accessibility and their relationship to land use was touched on in Chapter 1, in the section on urban land-use models. The **bid rent curve** (Figure 3.2a) is both a simple and an effective way of explaining variations of land use within cities and the concentration of commercial functions within the city centre. As commercial uses such as banking and insurance offices can afford the highest rents they are located within the CBD. As one moves away from the centre, the functions change in accordance to their need for accessibility and their ability to pay higher rents. As cities grow, their CBDs also grow and move outwards into the zone in transition and the inner suburbs. CBDs may indeed grow in many different ways; the case study of Barcelona below illustrates this point well.

Even within CBDs (especially in very large cities) there tend to be hierarchies of function according to the ability to pay rents. Whereas the very central part of a CBD, the **Peak Land Value Intersection (PLVI)**, where land values are at their highest, has the biggest concentration of banks, insurance offices, commodity exchanges and company headquarters, the major stores and retail facilities are often

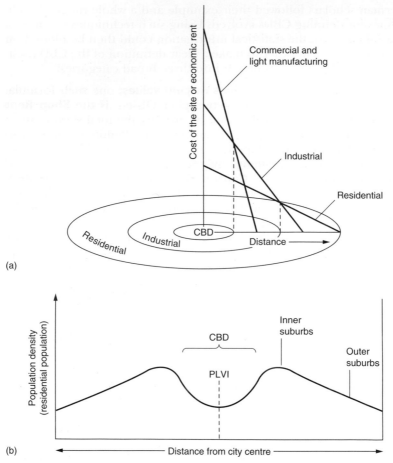

(a)

(b)

Figure 3.2 (a) The bid rent curve. (b) Urban population density graph.

located further away from it as they may not be able to afford such high rents. Generally, beyond this is another ring of CBD functions that take up more space and cannot afford such high rents as at the PLVI, yet still require centrality, and this includes such land uses as transport terminals, newspaper offices and broadcasting studios.

In contrast to the bid rent curve graph with its peak in the centre, population density within a large city (particularly within MEDC cities) has a very different sort of curve (see Figure 3.2b). The processes that have caused the highest bidding commercial functions to locate centrally have also forced a great proportion of former residential land use out of the CBD. Population density is therefore at its greatest in the inner city and dips within the CBD; this is known as the **'urban doughnut' effect.**

CASE STUDY: THE ONGOING EVOLUTION OF BARCELONA'S CBD

Barcelona, the capital of the Spanish autonomous region of Catalunya (Catalonia), with a population of 4.5 million within its built-up area, is Spain's second largest city. A major port, centre of commerce and industry, and an increasingly important tourist destination, Barcelona has a buoyant and dynamic economy that is reflected in the way in which its CBD has evolved through time and in how it is now expanding in new directions. Figure 3.3 shows the main structural elements of Barcelona's CBD.

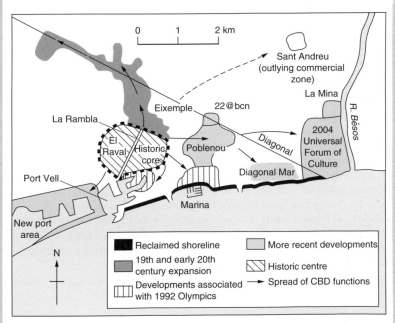

Figure 3.3 The ongoing evolution of Barcelona's CBD

Until the late eighteenth century, the city was contained within the Mediaeval walls and the majority of the commercial activity was concentrated there and at the adjoining port (Port Vell). In the mid-nineteenth century, a new elegant Garden City was planned by the architect Cérda to the north of the old city. This zone, Eixemple, is laid out on a rigid grid pattern of several hundred blocks and is cut through by one diagonal boulevard, La Diagonal. Most of the important commercial functions moved from the old city to Eixemple and are still concentrated there today, and include the main banks, insurance offices, travel agents, department stores, luxury goods shops, high-quality restaurants, large hotels and nightclubs.

The run up to the 1992 Olympic Games gave Barcelona a golden opportunity to carry out urban regeneration. This took place at an old industrial strip along the coast, where new beaches, parks, blocks of flats, restaurants, shops and other commercial facilities were built on brownfield sites, effectively extending the CBD eastwards. The next phases of Barcelona's commercial development came from the late 1990s onwards with the further reclamation of brownfield sites in the east. In the Poblenou area of the city, the high-tech area known as 22@bcn and the Diagonal Mar districts were completed for the 2004 Universal Forum of Cultures exhibition. Reclaiming old industrial land in the lower Bésos valley, the Diagonal Mar has a zoo, beaches, parks, 10 000 new apartments, hypermarkets, a shopping centre with 250 shops, a great deal of new office space and an exhibition centre. This development is also giving new life to one of the poorest suburbs, La Mina.

Back in the old city, rapid change is also taking place. The eastern part of the walled city, the Barri Gotic, has gradually become gentrified, with its many small hotels, bars and restaurants geared to the tourist industry, its museums, craft workshops and antique shops. The other side of La Rambla, the wide pedestrianised street that bisects the old city, is El Raval. This is a run-down maze of streets with a high immigrant population and cheap, but poor-quality accommodation. It has a wide range of social problems such as high unemployment, high crime rates, drug trafficking and prostitution. Now action is being taken to upgrade this area with developments such as a new contemporary art museum and university buildings.

4 Different Types of CBD

Although each CBD is unique, some general patterns can be seen which occur in different parts of the world. Some of the variations are explained by cultural factors, others by historical factors, and perhaps most important of all is the size of the city and therefore its field of influence. The bigger the city, the greater its influence and the wider its range of functions, the more likely it is to have a complex CBD.

a) The pre-industrial city

Pre-industrial cities have relatively small CBDs and by far the most numerous commercial functions are connected with retailing and other forms of trading. Open air and covered markets, shops, warehouses and small-scale workshop industries are all concentrated within the commercial district. One of the most striking features of these CBDs is the **clustering of functions**. In Mediaeval European cities and pre-industrial cities in LEDCs today, individual districts of the central core specialised in certain trades. Shoppers could therefore walk along a specific street and compare prices; these streets would have been the origin of the concept of **comparison stores**. In many European cities the names of trades and wares remain preserved in street names. A study of historic English town centres provides names such as Butter Market, Cloth Market, Corn Street, Dyer Street, Fish Street, Haymarket, Mercery Lane, Milk Street, Sadler Street, Silver Street, Weavers Lane and Wood Street.

Some of the best examples of pre-industrial cities in LEDCs can be seen in North Africa. The *medina* areas of cities such as Sousse, Sfax and Kairouan in Tunisia retain whole street complexes within their covered *souks* (markets) devoted to leatherworking, textiles, carpets, metalworking, jewellery and so on. When French colonial CBDs were added to their urban fabric in the nineteenth century, the old *medinas* remained largely unaffected.

b) The core and frame concept

Many medium to large size cities (with populations in the 1–2 million range) have developed a **core and frame** type of CBD. This has two distinctive parts to the commercial district: an inner core where the building density is greatest, reflecting higher land values, and an outer frame which has lower density buildings, lower rental values, yet still has functions which need to be centrally located. Victorian cities with grid or near-grid street patterns have often developed into this form. Manchester and Melbourne are two striking examples of the core and frame structure. In the case of Manchester, land use in the core includes the main concentrations of shops, incorporating the Arndale Centre, high-rise office blocks, the new Urbis centre (a museum celebrating world cities), the town hall, the central library, the cathedral and the restaurants of Chinatown. By contrast, the frame contains many functions which are more extensive forms of land use, including Victoria and Piccadilly railway stations, the Museum of Science and Technology, BBC North and Granada broadcasting studios, the University of Manchester Institute of Science and Technology, and the bars and restaurants of Canal Street.

c) The complex CBDs of world cities

The most important cities in the world have much more complexity in the structure of their CBDs, and can be regarded as being **polycentric**. London, New York and Tokyo are the supreme examples of CBD complexity. What happens in all three cases is that different

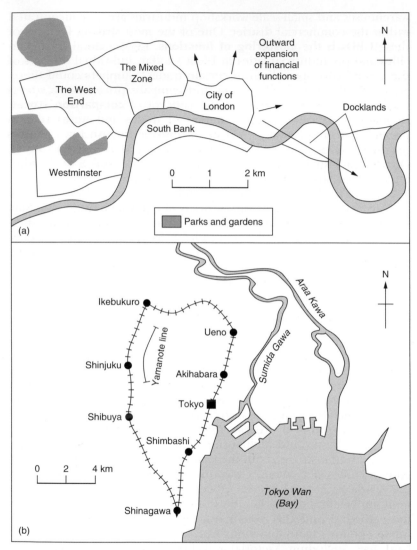

Figure 3.4 The CBDs of London and Tokyo: (a) the main areas of London's CBD, (b) Tokyo's nodal points along the Yamanote line

types of function have become clustered into different parts of the central area. This is found in many large cities throughout the world, particularly major capital and port cities where the administrative and import and export functions are likely to be separate and distinct in character. However, London, New York and Tokyo offer the most interesting examples of complex CBDs and two of these, London and Tokyo (see Figure 3.4), will be examined in detail.

i) London

London already had a complex core before the Industrial Revolution and its rapid eighteenth- and nineteenth-century expansion. The old Roman core coincided with the area that is the City of London today, the main financial district. With the fall of the Roman Empire in the fifth century CE, there was a westwards shift of the built-up area towards Ludenwic, centred around what is the Covent Garden area today. Another, separate settlement was created upstream at Thorney Island (today's Westminster) in Saxon times and this became the main centre of political, rather than commercial power by the Mediaeval period. Looking at the map of London's CBD (Figure 3.4a), it can be seen that there are seven main zones, some central, others not.

Today, the City of London with the Bank of England, hundreds of foreign bank headquarters, Lloyd's insurance house, the Stock Exchange and LIFFE (the London International Financial Futures Exchange) is still the main financial district of London's CBD. By contrast, the West End has London's main concentration of retailing, cultural and entertainment facilities, including the department stores of Oxford and Bond Streets, museums and galleries such as the National Gallery, the majority of London's theatres, and the large concentration of restaurants, bars and nightclubs, such as in the Soho district.

Between the City and the West End lies an area of mixed land use in the Holborn and Bloomsbury districts, dominated by the legal profession (the Inns of Court) and constituent colleges of London University (King's College, University College and the London School of Economics), as well as the British Museum and the 1980s' redevelopment of Covent Garden Market and its neighbouring pedestrianised streets into an important centre of retailing and entertainment.

Westminster was established as the centre of government in the Middle Ages and has remained so, with Parliament and ministries, as well as royal and prime ministerial residences there today.

From the 1950s onwards, and still continuing today, the South Bank of the Thames has been transformed from a riverside industrial site to one with mixed central city functions; these include the South Bank cultural complex (including the Royal Festival Hall and National Theatre), the Butler's Wharf complex of shops, restaurants and offices, the new Greater London Authority Building and the reconstructed Globe Theatre.

From the 1960s onwards, when office devolution was taking place in London, huge new office blocks were built on the southern edge of London at Croydon where rents are much lower than in the city centre and a large number of companies and parts of the civil service moved there. Thus, Croydon has become effectively a detached or out-of-town part of the CBD. Similarly, from the 1990s, London Docklands, which now has the appearance of a 'mini-Manhattan', has been the focus of relocation from the city centre. This part of London's CBD has attracted bank headquarters away from the City in particular. HSBC, Citigroup and Barclay's are now all based at Canary Wharf.

ii) Tokyo

Although the site of central Tokyo is predominantly flat, including the flood plain of the Ara Kawa and Sumida Gawa, there is a series of low hills on which historic settlements were established. Today, these locations are linked by the Yamanote Line, a roughly circular overground railway line of the Tokyo metro, and they form the complex CBD of the city (Figure 3.4b). Although each of the parts of the CBD has commercial functions such as shopping complexes, banks and restaurants, as in London, each has a degree of specialisation. The settlement named Tokyo is the most important historic location as it is the site of the Imperial Palace and park, and it also has the main central railway station and several of the most important department stores in the adjacent Ginza district.

Shinjuku is the most important business district with banks and other financial institutions, the city's biggest department stores such as Isetan and Takashimaja, as well as having the biggest concentration of nightclubs (including the red-light district), major high-rise hotels, restaurants and the city's twin-towered town hall. Shibuya, likewise, is a centre for shopping and entertainment, but also has the NHK broadcasting complex, sporting facilities from the 1964 Olympics, and Tokyo's most important Shinto religious building, the Meiji Shrine, set in its gardens. Ueno is a major route centre for the railways and important open-air market, but is also where the city's main museums and concert halls are located. Akihabara is based around Tokyo's largest Buddhist temple and market complex, but is also the centre of the electronics and high-tech retailing zone. Harujuku is the centre of the fashion industry, with a large concentration of boutiques and small stores appealing to the young, and includes the Takeshite pedestrianised street.

5 Public Space and Private Space

One of the greatest concerns in the contemporary city is the encroachment of private ownership into spaces that were traditionally regarded as public. The traditional market place is where people meet at different times of day for a diverse range of purposes, from buying and selling to entertainment and public processions and protests. The square, *plaza*, *piazza* and its equivalents in other countries and cultures is in danger of being replaced by private spaces, where activities are much more restricted.

In North America, a new model has been established which is being emulated throughout the world. The shopping and entertainment mall is becoming the new climate-controlled indoor equivalent of the market place in many cities. There are many interrelated reasons for this, including:

- the decline and decay of many downtown areas and the consequent flight of populations to the suburbs
- ever-increasing growth of car ownership and the increased mobility that comes with it
- extreme summer and winter climates of many North American cities
- commercial opportunism of large, speculative property development companies and shopping chains in creating new lifestyle cultures and retail habits.

Unlike traditional market places, shopping malls have restricted opening hours, private guards who scrutinise people as they enter the space, widespread surveillance camera coverage and private enterprise governing all activities within the complex. Effectively, the shopping and entertainment mall represents a form of privatisation of public space. In many places, malls are taking themed forms. Perhaps an indoor, air-conditioned pastiche of a Mediaeval European market square will be built somewhere soon.

6 The Conservation of Central Heritage Areas

City centres generally have the highest concentrations of historic buildings. The way in which these are treated and evaluated depends greatly on the way they are perceived by a particular cultural group or generation. In an age in which tourism has become the most important source of income to many countries, heritage is often being considered more carefully by city and national governments than it was in the past. For example, in former Communist countries religious buildings are being restored and cared for more than they were two decades ago. Newly industrialised countries (NICs) are also guarding their architectural heritage in a way not seen until quite recently; the case study below will look at one such example, Singapore.

In Britain, a great deal of damage was done to historic city centres in the 1960s and 1970s with the building of inner ring roads and the redevelopment of shopping areas. Even a city of such great architectural importance as Bath suffered from demolition of the old and the building of characterless new structures such as multi-storey car parks and a new bus station and shopping area in the south of its CBD. Lessons have been learned from the past and Bath's central conservation area has been restored to a level expected of a cultural site of world importance. In establishing its World Heritage Site List, UNESCO (The United Nations Educational, Scientific and Cultural

Organisation) has achieved a great deal for individual historic cities (Bath joined the list in 1987). Places as diverse as the abandoned cave-dwelling city of Matera in southern Italy (1995), which is partially being rehabilitated and recolonised, and the walled Spanish colonial city of Campeche in Mexico (1999) have benefited greatly from their UNESCO status both in their restoration and in the subsequent focus of tourism.

CASE STUDY: SINGAPORE

Singapore put little emphasis on the conservation of its everyday architecture until the 1980s. There were numerous reasons for this, including the expansion of the high-rise CBD into the historic areas, the clearing of old residential buildings, which were often seen as a health risk for the widespread public housing schemes, and the attitude of the nationalist government towards the colonial past. In the 1970s many large, important individual buildings were being listed as national monuments (e.g. churches and temples), at the same time as many commercial buildings and in particular rows of characteristic 'shophouses' (a building with a shop on the ground floor and accommodation for the shopkeeper on the first floor) were being demolished. By the mid-1980s there was a change in direction when the government realised that whole streets of houses in the main ethnic areas of the old city were rapidly disappearing; their preservation was seen as important both for Singaporeans as symbols of national identity and for foreigners as tourist attractions.

As a result of the report of the Urban Redevelopment Authority in 1984, four major conservation areas were established: the historic Civic District (with many British colonial buildings), Chinatown (representing the traditional architecture of Singapore's ethnic majority), Little India and Kampong Glam (representing the most significant historic areas occupied by the Indian and Malay ethnic minorities). Areas that could easily have been lost are now major selling points of the Singapore tourist industry. Chinatown, in particular, is now attracting a wide range of commercial activities. It retains a very large concentration of shops and restaurants, but at the same time many of its old shophouses have been converted into prestigious offices for the professions and arts and media-based enterprises, as well as bars and clubs which are an important part of Singapore's rapidly expanding nightlife scene.

7 Decentralisation and the Devolution of CBD Functions

The late nineteenth century saw the great concentration of banks, offices and other commercial buildings in city centres. By the mid-twentieth century this concentrated form of land use was being regarded in many cities as a liability rather than an asset. Traffic congestion, longer journeys to work and burgeoning rental values led city leaders to reassess the wisdom of having so many commercial and administrative functions crammed within CBDs. This was generally most marked in cities such as London, Tokyo and Paris that are, at the same time, capital cities and major commercial centres to their countries.

From the 1960s onwards, London has experienced trickles and flows of employees being relocated elsewhere in both the public and private sectors. In the last few decades, within the civil service, most departments have established regional offices with personnel moved from London. At the same time, whole government offices have relocated, e.g. the Manpower Services Commission to Sheffield, the Export Credits Guarantee Department to Cardiff, Her Majesty's Stationery Office to Norwich and the Small Industries in Rural Areas office to Salisbury. In the 1970s and 1980s Yorkshire, the North West and Scotland received the largest number of devolved offices when these regions had high unemployment rates. Since the 1990s, the South West has been the main area for relocation. Examples of private sector relocations of headquarters from London include the AA (Automobile Association) to Basingstoke, WHSmith to Swindon and Guardian Royal Exchange Assurance to Ipswich. In all cases the organisations benefit from lower costs of living and consequent lower salary bills to be paid, as well as lower rental values.

Devolution of central functions also takes place within city boundaries; as mentioned above many of London's commercial offices were relocated to Croydon in the 1960s and 1970s and then to the Docklands in the late 1980s and 1990s. Similar relocations have occurred in other European cities. In Paris, the western corridor formed by La Defence became an area to which a wide range of central functions was relocated in the 1980s. In Rome, a similar project will be taking place in the next decade and this is considered below.

CASE STUDY: ROME AND ITS SDO

When Rome became the capital of the new united Italy in 1871, it had greatly shrunk in size since ancient times and there was a lot of open space for development within the third-century Aurelian walls. There was rapid change in the next few decades as new ministries were built, and the headquarters of hundreds of other institutions followed. The city grew from a population of 250 000 in 1871 to just under 1 million by 1930. Then, in the post-war decades of economic boom, Rome experienced even greater growth until it reached its population peak of 2.8 million in 1981.

Although not an industrial city, Rome has a wide range of tertiary activities, which are concentrated in the historic core. As well as the administration of four levels of government (national, regional, provincial and urban), there are two sets of embassies (to Italy and to the Vatican), several agencies of the United Nations, hundreds of company headquarters and hundreds of hotels to accommodate the 5 million tourists who descend on the city every year.

The historic core is dominated by a dense network of narrow Mediaeval and later streets, and this leads to huge traffic congestion problems at rush hours. The historic nature of most of the administrative buildings often makes them ill-equipped for the requirements of modern technology.

In the boom years after the Second World War, some government functions and company headquarters (such as the state hydrocarbons company, IP) moved their offices into the suburb of the EUR district (*Esposizione Universale di Roma*). This was to have been the location of a trade fair in 1942, but war intervened. Several of the national museums were also located in this well-planned garden suburb to the south of the CBD. Other than this, the situation in the historic centre, with increasing car numbers and a mushrooming of tertiary functions, progressively became ever more severe. In the city's strategic plan of 1962, proposals were made to move much of the government bureaucracy out to a series of purpose-built sites in the eastern suburbs of Rome – the SDO (*Sistema Direzionale Orientale*, meaning the 'eastern directional system of offices'). The SDO would also include housing for 225 000 people and a wide range of services, and would be linked to the city's traffic network by fast new roads and metro lines.

The proposal was debated for decades and in 1995 was abandoned on grounds of cost and the reluctance of people to work outside the city centre. In 2003, interest in the plan was revived and work started on the first phase of development in 2005. With the investment of €161 million, work began on a 130 hectare site at Pietralata, one of the three main suburban nodal points in the eastern suburbs where the SDO will be located (see Figure 3.5). The office buildings are intended to have a high degree of sustainability in the way in which they operate, as well as a flexibility to enable them to be adapted to future needs. The provision of the right level of transport infrastructure in time for the inaugu-

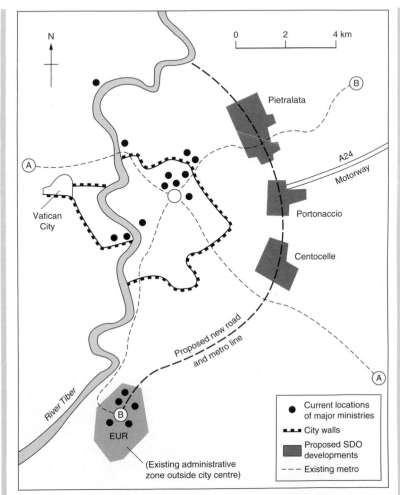

Figure 3.5 The location of Rome's SDO

ration of the SDO is very important; the lack of good transport links was one of the main reasons why London's Canary Wharf development was slow to become a success.

8 Recentralisation

In many patterns of human geography, one process is followed by a counter-process. In many MEDCs, particularly in the USA during the second half of the twentieth century, CBDs were allowed through the 'natural' forces of capitalism to become rundown and to some degree abandoned by both the service industries and the middle classes. Problems such as dereliction of land, crime and racial tension have haunted parts of city centres for decades. The closing down of industries, warehouses and transport facilities in the zones in transition (ZITs) immediately around CBDs, as a part of the de-industrialisation process, further isolated central areas from the rest of the economically active parts of cities. Revitalisation schemes have been put into action to 'reclaim' the centres for the use of the general public and to attract back the functions normally associated with CBDs but which had been devolved into more wealthy suburban and city edge locations. Los Angeles presents perhaps the most dramatic example of this in the USA.

CASE STUDY: 'DOWNTOWN' – THE CBD OF LOS ANGELES

Los Angeles has already been discussed as the model of the 'post-modern' or even 'post-metropolitan' city, yet even in that sprawling Californian megacity, the importance of the CBD has been re-assessed and a huge amount of money was invested in

The revitalised CBD of Los Angeles, USA

revitalising it in the decade from 1995 to 2005. This regeneration coincided with Roberta Brandes Gratz's 1998 book *Cities Back from the Edge*, which recognised similar trends in other US cities. By the 1980s much of central Los Angeles had become so run down and dominated by its immigrant Hispanic community that huge numbers of wealthier people had moved out to the suburbs, as had many of the main financial institutions.

The organisations responsible for Los Angeles' regeneration designated LADCBID (the Los Angeles Downtown Centre Business District), a 60-block area with over 400 properties. Institutions involved in this regeneration include the LA County Economic Development Corporation (LACEDC) and the Central City Association (CCA). The results of their investments and other initiatives are clearly visible with the development of a cluster of new high-rise office blocks (with a restored funicular railway linking these hilltop blocks to the streets below), the redevelopment of the area around the Union Station including the Hispanic 'Old Plaza' with its open-air stalls and craft market, the construction of new hotels and the opening of new museums and cultural spaces. As well as the 28-storey City Hall which was the only high-rise in Los Angeles when built in 1928 and which was reinforced in 1996 to become more earthquake resistant, Downtown now has two other iconic buildings: the new Roman Catholic cathedral of Our Lady of the Angels, completed in 2002, and the Walt Disney Concert Hall, designed by Frank Gehry and inaugurated in 2004. Although the wholesale trade remains the biggest employer in the Downtown area (27%), professional, technical and financial services are now the second most important (17%), followed by the more traditional manufacturing and retailing sectors (16% and 11%, respectively). The financial and technical services are destined to become the most important form of land use in the near future.

Questions

1. Explain how land values and competition for space have determined the evolution and nature of contemporary CBDs.
2. With reference to specific examples, explain why the CBDs of cities of different sizes have very different characteristics.
3. Outline the main causes and effects of the devolution of CBD functions.
4. With reference to specific examples discuss why city centres are becoming more important as tourist destinations.

4 The Inner City and the Zone in Transition

'Cities are scary and impersonal, and the best most of us can manage is our fragile hold on our route through the streets … . The freedom of the city is enormous. Here one can choose and invent one's society, and live more deliberately than anywhere else. Nothing is fixed, the possibilities are endless.'

Jonathan Raban *Soft City*

Nowhere in the city does the apparently paradoxical observation of Raban hold more true than in that most decaying yet most dynamic zone found immediately around the CBD. The term **zone in transition (ZIT)** has been widely used since it appeared as part of the Burgess Model (see page 18) in 1922. Traditionally an area with distinctive forms of land use related to both the CBD and inner suburbs, the ZIT in many cities is undergoing even more dynamic change than the city centre. There are two main reasons for this:

- as has always been the case, parts of the ZIT are receiving an increasing number of CBD-related functions as the CBD expands outwards owing to shortages of space
- in former industrial cities, there are vast areas of obsolete plant and infrastructure, dating back to the nineteenth century, located within these older inner city areas.

1 The Definition and Development of the ZIT

The ZIT is, by its very nature, an area that is rather fluid and difficult to define spatially. It has an ever-growing percentage of commercial functions and a declining proportion of traditional functions such as workshop industries, warehouses and transport land. Where redevelopments of a particular type are taking place, such as so-called 'loft conversions', ZITs are also increasing their residential capacity. By measuring the proportions of these three types of land use, the rate of change within a ZIT can be established. However, urban geographers have not come up with any magic formulae that can define exactly when an area becomes a ZIT, or when a ZIT is sufficiently commercial in function to become *de facto* a part of the CBD. The way in which a ZIT area undergoes its transition is best illustrated by the Clerkenwell case study below.

2 The Regeneration of Inner City Areas

No other part of any large urban area has given greater opportunities for **regeneration** in the past few decades than the inner city/zone in transition. CBDs are constantly undergoing change, and in this respect have tended to keep up with the times technologically and architecturally. High land values and the competition for space have meant that only small areas of CBDs have been left vacant at any one time.

This has not been true of ZIT, where much larger areas of derelict land may have been left unattended for decades or under extensive forms of land use not normally associated with modern central city areas, e.g. storage depots, motor garages and workshops, and rubbish recycling units. These areas, the product of de-industrialisation and the consequent shift of urban economies away from centrally located heavy industries towards the tertiary sector and to lighter industries located in the suburbs, have given cities their greatest generation potential in the late 1990s and early 2000s. These run-down and derelict inner urban areas can be put into three main categories:

- those associated with former manufacturing industries (workshop or factory based)
- those associated with ports, harbours, riverside or canalside locations
- those associated with railway land.

The first major inner city regeneration schemes were concentrated mainly in the USA in the 1970s and early 1980s. At this time British cities were still largely concentrating on CBD redevelopments and housing projects and therefore neglecting their ZITs. In the USA some of the first major projects were in Baltimore (Inner Harbor) and in New York (Seaport South).

a) The regeneration of old industrial areas

Most British cities once had substantial industrial areas on the fringes of their CBDs that dated back to the earlier phases of the Industrial Revolution. These industries were often closely associated with either the activities of the CBD, e.g. printing, or the large market of wealthy city workers near at hand, e.g. jewellery making.

CASE STUDY: DE-INDUSTRIALISATION AND CHANGE IN CLERKENWELL, LONDON

In London, the Clerkenwell and Finsbury areas immediately to the north of the City provide a striking example of such an area that has undergone dramatic changes in the past 15 years. From the mid-nineteenth century it was an industrial area, generally operating on a small-scale workshop basis, but with some larger factory units. Some of its main works included breweries, glassworks, lead and other non-ferrous metal producers, printworks, jewellery and precision instrument makers, with its greatest specialisation in clock and watch manufacturers. As late as the 1980s St John's Street, the main north–south artery through Clerkenwell, had over a dozen watchmakers, by 1992 this had fallen to just three and in 2004 there was just one watch repair shop left, sharing its premises with a solicitor. With de-industrialisation a wide variety of functions have moved into the area, including:

- high-quality bars and restaurants aimed at the professional workers
- new-style offices with high design specifications, particularly attracting media and advertising industries
- designer shops (e.g. for high-quality office furniture)
- photographic laboratories and digital printers (reflecting some continuity from the older functions of the area).

These changes have given the area a vitality and 'buzz' that it lacked right from the end of the Second World War until the late 1990s. The area is no longer totally dead at night as many of the bars and restaurants are magnets to young and professional people in the evening. Clerkenwell has become one of the most sought-after places to live in London and a great deal of old industrial and warehouse space has been converted into residential functions. One of the advantages of occupying such spaces is that they are often open plan and allow owners to design their own home layouts.

On the negative side, one of the high-rise blocks of local authority housing was demolished in the late 1990s and it has been replaced by luxury housing. Social changes are therefore taking place and some of the more established families are being forced out of the area by a lack of job opportunities and rising property costs. In January 2005 an article in a London newspaper, the *Evening Standard*, considered the changes taking place in St John's Street and highlighted the fact that it now has over 30 bars and restaurants, most of which opened in the previous five years. The typical clientele of these venues are City workers in their thirties and early forties who can afford to pay £50–80 for a meal. Coining the term the 'Clerkenwell Effect', the newspaper also considered Exmouth Market in neighbouring Finsbury as a street that has undergone similar changes in recent years. With the lengthening of opening hours of licensed premises in 2005, some parts of Clerkenwell and Finsbury may become hot spots for noise and nuisance to the local residents in the small hours of the morning.

b) The regeneration of old port, riverside and canalside areas

The process of de-industrialisation, together with the development of ever larger ships and the containerisation of shipping freight, led between the 1970s and the 1990s to the abandonment of old ports and the creation of new ones. In the case of London this meant the migration of port facilities downstream from the East End to places such as Tilbury and Gravesend. In Bristol, there was a similar shift from the city centre to Avonmouth on the Severn estuary; on Tyneside, similarly there has been a shift from central Newcastle and Gateshead to further down the Tyne. Throughout the world, river port cities have had to build new container ports to replace their obsolete, more central facilities in order to remain competitive, and this is as much the situation in some NIC cities, such as Singapore, as it is in MEDC cities.

Certain industrial cities relied more heavily on canals than rivers for their transporting of goods. With changes in shipping, canals have become largely obsolete in Britain for freight transport since 1950, and although there are moves to revive canals for goods traffic, canalside areas in large industrial cities have been redeveloped for other types of land use. In London, the King's Cross Basin on the Regent's Canal has been redeveloped for luxury housing and a Canal Museum. The Salford Quays project in Greater Manchester has been developed on the basins of the Manchester Ship Canal and includes buildings such as The Lowry arts complex, the Imperial War Museum North and a variety of sports infrastructure projects for the 2002 Commonwealth Games, as well as large areas of housing on different scales.

CASE STUDY: REGENERATION ALONG THE SINGAPORE RIVER

One of the most remarkable recent transformations of riverside and old portside areas is the ongoing changes that are taking place in Singapore (see Figure 4.1). Close to the centre of the city is the Singapore River. This river was used for local freight traffic when it was off-loaded from the main port and taken to the city centre; the size of vessels was small but the volume of traffic was great. With the closing of many commercial properties along the riverside and the transfer of most local freight transport from water to road, the Singapore River was ripe for development. The original plans in the 1970s were to demolish the shophouses lining the river and to include the land they used into part of the high-rise CBD. However, a much more imaginative approach was followed and in the mid-1980s the shophouses were converted into what is today the Boat Quay complex. The low-rise traditional buildings that have the CBD towering behind them are today a row of vibrant bars and restaurants patronised by city workers and tourists alike. In the Singapore strategic plan of 1992, this and other areas along the Singapore River were called 'The Night Zone' or 'The City that Never Sleeps'.

Further upstream is Clarke Quay, developed from mainly warehouses in the mid-1990s. This development as yet lacks the patronage and success of Boat Quay, partly owing to its location further away from the CBD and potential customers and partly owing to its less authentic feel. The conversion of the warehouses

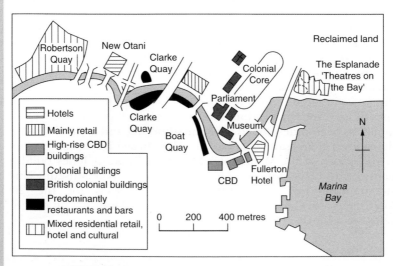

Figure 4.1 Development along the Singapore River

Regeneration along the Singapore River

into bars, restaurants and shops provides a less attractive environment, and some of its functions are questionable, such as the flea market that has been planned and lacks spontaneity. Further upstream still is the Robinson Quay complex built in the late 1990s. This is mainly an area of condominium housing for the more wealthy Singaporeans, although certain arts functions such as the Singapore Repertory Theatre have been included within it. Robinson Quay has been criticised as being a rather bland development along what is a rather sterile and lifeless river, and as with so many inner city enterprises throughout the world it has replaced a popular and lively local population with a rich middle class one.

At the mouth of the Singapore River on reclaimed land is one of the city state's greatest recent developments, the Esplanade. This concert hall and theatre complex opened in 2003 and has given Singapore some iconic new architecture.

c) The regeneration of old railway land

Railways have traditionally taken up large areas of land within cities, and often close to city centres. In addition to stations themselves, sidings, marshalling yards, coal yards, warehouses and other storage spaces were part of urban railway infrastructure. Changing technology, together with increased pressure on land and its value, have led to the release of a considerable amount of railway land in recent decades in many countries. In Britain, it was the privatisation of railways which, in particular, put profit making centre stage and this has led to the development of a lot of former railway lands.

CASE STUDY: THE REGENERATION OF THE KING'S CROSS AREA, LONDON

The railway land just to the north of King's Cross station is one of the largest undeveloped ZIT areas with easy access to the centre of London. Most of the land and the buildings on it have remained in a state of dereliction or semi-dereliction for several decades. There have been conflicts of interest between the railway companies and the local councils, which have held up the decision-making processes as to how the land should be developed.

The breakthrough came in the late 1980s when the decision was made to use the land between the existing King's Cross and St Pancras stations as the location for the newly projected London terminal of the Channel Tunnel Rail Link. In the early 1990s a consortium called the London Regeneration Corporation came up with a master plan for redeveloping the site. The new station was to be built on a wedge-shape area occupied by the Great Northern Hotel, which along with many of the other railway buildings would have to be demolished. A new city of offices, luxury housing, parks, shops and entertainment facilities was to have been built. The proposed layout had an almost Renaissance geometric pattern to it. However, by the late 1990s the plan had flopped, for a variety of reasons:

- for reasons of cost, the idea of building a new station was dropped and the cheaper option of using a revitalised St Pancras Station as the Channel Tunnel Terminal was adopted
- both Camden and Islington Councils (in whose boroughs the land lies) objected to the large proportion of commercial land use and the small amount of low-cost housing
- local conservation groups objected to the proposed uprooting and relocation of the mature ecosystem of the Camley Street Nature Reserve along the Regent's Canal.

In the late 1990s a compromise solution evolved which satisfied the local councils was started and projected to be complete by 2007 when trains from Paris and Brussels are scheduled to arrive at St Pancras (see Figure 4.2). The first two projects connected with the scheme have been at St Pancras Station and the new Regent Quarter development. At St Pancras the platforms were extended to take the Eurostar trains, the station itself was upgraded and the Midland Hotel – which had been semi-derelict for decades and under threat of demolition – was being refurbished as a luxury hotel with penthouse apartments.

The Regent Quarter development is having a profound effect on the area. It is replacing a seedy, run-down zone associated in

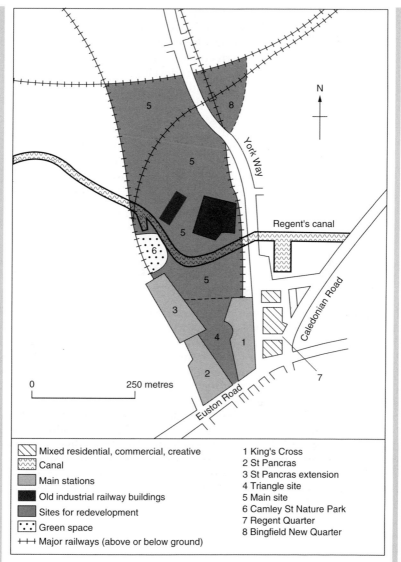

Figure 4.2 King's Cross redevelopment in London

the past with crime, drug dealing and prostitution. Taking its name from the nearby Règent's Canal and being developed by P&O Properties, most of the quarter is comprised of refurbished warehouses and industrial buildings, together with modern infill. The new land use of the quarter is to be mainly commercial and cultural, with a large hotel, many restaurants and bars, shops, a gallery, a dance studio and a theatre. It is part of a bigger redevelopment area called the King's Cross Creative Industries Quarter. This is one of nine neighbourhoods in Islington that border King's Cross and are being upgraded through Area Action Plans (AAPs) through consultation with local residents.

The two major areas due for redevelopment behind King's Cross Station are the Triangle and the Main Site. These cover 27 hectares and plans have been put forward by a consortium of three firms: Argent St George, London Continental Railways (the main land owner) and Exel. The Triangle Site is to be a piazza known as Station Square. The Main Site will retain the Camley Street Nature Reserve and some of the more historic buildings and will include offices, shops, housing of a variety of types and some recreational spaces. If plans are approved construction will start in 2007.

3 The Spread of High-rise CBD Buildings into Inner City Areas

It is almost inevitable that as cities and their service industries expand there is increasing pressure on space in city centres and their immediate peripheries. In a large number of cities, in both MEDCs and LEDCs, high-rise blocks, often with mixed commercial and residential functions, are therefore now being built in inner city areas. In some instances such as at Canary Wharf in London and the South Bank development on the other side of the Yarra River from the city centre of Melbourne there has been considerable clustering of high-rise buildings, creating 'mini-Manhattan'-type skylines. In other places there are more controversial proposals that would involve one high-rise building with other new low-rise buildings clustered around it.

In London there are at present many proposals in various stages of consideration that would take high-rise into ZIT areas. Following the events of 11 September 2001, there was a lot of opposition to the construction of any more high-rise blocks in central London. Another reason why new high-rise in London was criticised was the potential destruction of viewpoints, in particular views of historic buildings such as St Paul's Cathedral. London's mayor has been in favour of having more tall buildings in the capital, and the success of the SwissRe Tower ('The Gherkin') as a London landmark has rather turned public opinion.

Parts of London that are destined to have high-rise commercial and residential buildings added to their skylines (if planning permission is granted) include the Paddington Basin, Elephant & Castle and London Bridge Station. At 310 metres, the so-called 'Shard of Glass' at London Bridge would become the tallest building in the capital.

4 Inner City Entertainment and Culture Zones

In the post-industrial age, one of the most important changes of function within the ZIT and other inner city areas has been the growth of leisure and recreation. The presence of old industrial buildings such as warehouses, workshops, power stations, wholesale markets, factories and railway arches has offered great possibilities for the development of leisure facilities such as bars, nightclubs, restaurants, fitness centres, performance spaces and art galleries.

In London, the Clerkenwell area has been mentioned in a case study above. Other inner city areas that have been adapted to leisure functions include several of the large former warehouse buildings behind King's Cross Station which are used as mega-nightclubs, as are various spaces under railway arches in such locations as London Bridge and Vauxhall. The old industrial buildings of the Hoxton area of Hackney have been transformed into artists' studios and gallery spaces, as well as bars, clubs and restaurants.

Whole inner city suburbs develop into attractive locations for the young and the more bohemian elements in society because of the sheer concentration of these leisure facilities. In New York, the SoHo, Chelsea and Greenwich Village neighbourhoods that lie between the two high-rise commercial districts of Downtown and Midtown Manhattan fall into this category, as do districts such as Paddington in Sydney and Ponsonby in Auckland.

Questions

1. Outline the processes of **de-industrialisation** and the general effects they have had on the changes in land use in inner cities.
2. With reference to specific examples, discuss how old industrial land close to city centres is being transformed.
3. What are the possible land-use conflicts that may occur when inner city redevelopments occur?
4. Why are so many regeneration schemes in inner city areas more likely to favour the rich rather than the poor?

5 The Residential Environment

'Social segregation in cities is not new. Rich and poor have been geographically divided since the start of the Industrial Revolution. The advent of mass urbanisation and industrialisation during the nineteenth century saw the creation of working class slum areas on a scale hitherto unknown.'

Chris Hamnett *Unequal City*

1 The Origins of Residential Differentiation in Pre-industrial and Industrial Cities

The origins of the segregation of functions within cities were discussed in Chapter 1. In pre-industrial cities there were distinctive zones within urban areas, but they were not always as great as they are today. This is particularly true of residential functions. The ruling classes and wealthiest segments of society may have lived in a distinctive zone, as was common in certain parts of Europe from Mediaeval times, but this is not necessarily typical of all pre-industrial situations. The reliance of one social or economic group on others and the slow and limited nature of transport within cities led to the rich, the merchant class, craftsmen and manual workers all living in close proximity as there was a close interrelationship between their activities. This is still reflected in the social structure of residential areas today in cities as diverse as the compound towns of western Africa (e.g. Nigeria and Mali), the *medinas* of North Africa (e.g. Tunisia and Morocco) and historic cores of southern European cities (e.g. Italy and Spain). In such places, the palaces of the wealthy are located cheek-by-jowl with workshops, shops and the apartments of people from the full range of economic circumstances.

As Hamnett states in the quotation at the beginning of this chapter, the origins of today's more extreme form of residential segregation lie in the Industrial Revolution. The new economic system required a very large unskilled labour force who were accommodated in low-quality housing. Workers needed to live close to the factories because of their limited mobility. The industrial areas were noisy, overcrowded and polluted and the wealthier classes had new means of transport that enabled them to live away from city centres and these industrial zones. This was the beginning of the modern pattern of social polarisation in cities. The way in which this segregation took place is considered in the next section.

2 Variations in Residential Environments in MEDC Cities

The land-use models examined in Chapter 1 help to explain how residential variations evolve and how cities may continue to become socially more polarised. Several important factors lie behind the spatial arrangement of rich and poor neighbourhoods within cities. These are:

- topographical factors, such as hills, valleys, plains and rivers that have made certain parts of cities more environmentally desirable than others, and therefore those who can afford it will tend to live in those residential areas which are more scenically attractive
- factors relating to pollution and air quality that are often closely linked to the previous point; in the case of many British cities, the more expensive residential areas were historically located in the west as they would have been upwind of the industrial zones. This is far less important a factor in post-industrial times, but once patterns are fixed they continue on their momentum
- the laws of economics that led to social segregation in cities in the past, with the more wealthy choosing to live away from poorer areas; this too has largely continued on its own momentum
- the clustering of similar types of neighbourhoods in certain parts of the city that also resulted from these economic and social factors
- the dynamics of change that have in recent decades tended to make the patterns more complex as some areas become upgraded and others become downgraded; residential areas undergo changes in fortune which can lead to considerable migration into them or out of them.

Figure 5.1 conveys the complexity of the urban dynamics at work in MEDC cities today. Whereas it is the economic, occupational and

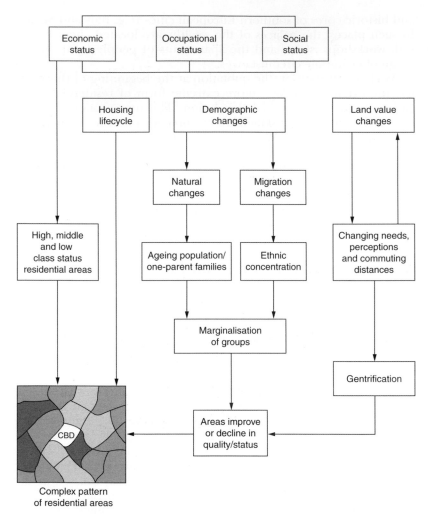

Figure 5.1 The dynamics of urban change in cities in MEDCs

social status of individuals and groups that may initially provide the fundamental framework for differing residential areas within cities, there are also the forces of the **housing lifecycle**, demographic processes and changing land values which provide the dynamic changes which lead to constant shifts in this residential mosaic.

3 The Residential Lifecycle

Residential lifecycles take two forms: those reflecting changes in economic status of families as they go through the chronological

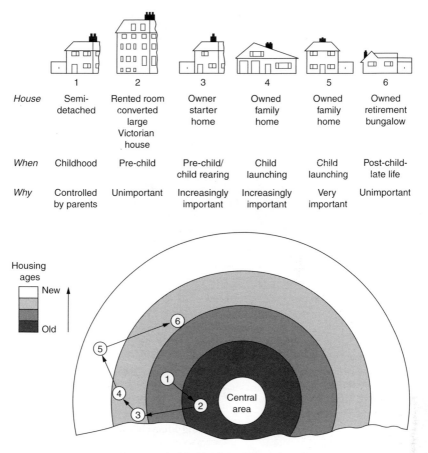

House	Semi-detached	Rented room converted large Victorian house	Owner starter home	Owned family home	Owned family home	Owned retirement bungalow
	1	2	3	4	5	6
When	Childhood	Pre-child	Pre-child/ child rearing	Child launching	Child launching	Post-child-late life
Why	Controlled by parents	Unimportant	Increasingly important	Increasingly important	Very important	Unimportant

Figure 5.2 The housing lifecycle – 1

phases of their lives, and those reflecting the changing fortunes of the buildings themselves. The sociologist John Rex devised the concept of the lifecycle of British housing in relation to different social groups in the 1970s. Figure 5.2 illustrates the lifecycle for a typical middle class family. It shows how, at different stages in its development, a family has different requirements in its accommodation needs and therefore makes a series of moves accordingly. The move typically takes the family outwards as it expands and its income increases, but then, as the children leave home, the family moves back towards the centre of the city.

Figure 5.3 is a chart reflecting the fortunes of a typical terraced house in a well-to-do British inner city suburb. It starts off its life as the family home with servants' quarters in the mid-nineteenth century. Its

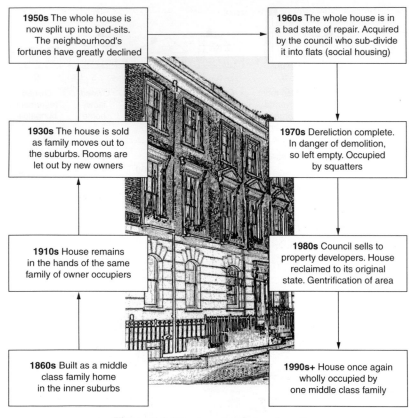

1950s The whole house is now split up into bed-sits. The neighbourhood's fortunes have greatly declined

1960s The whole house is in a bad state of repair. Acquired by the council who sub-divide it into flats (social housing)

1930s The house is sold as family moves out to the suburbs. Rooms are let out by new owners

1970s Dereliction complete. In danger of demolition, so left empty. Occupied by squatters

1910s House remains in the hands of the same family of owner occupiers

1980s Council sells to property developers. House reclaimed to its original state. Gentrification of area

1860s Built as a middle class family home in the inner suburbs

1990s+ House once again wholly occupied by one middle class family

Figure 5.3 The housing lifecycle – 2

fortunes gradually decline during the course of the twentieth century as it becomes subdivided into flats. The cycle reaches rock bottom in the 1970s, when it is semi-derelict, occupied by squatters and narrowly avoids demolition. From the 1980s onwards, however, it undergoes **gentrification** and becomes a family home once again. Gentrification is dealt with in more detail below.

Closely associated with these lifecycles is the process of **filtering** (see Figure 5.4). As the more socially and economically mobile families move outwards from inner suburbs to the low-density locations of the outer suburbs, the vacated housing in these inner suburbs is occupied by new migrant groups. In the case of British cities, the new groups are generally different waves of ethnic minorities. Ultimately, however, the new groups moving in will be the wealthier middle class professionals, the gentrifiers.

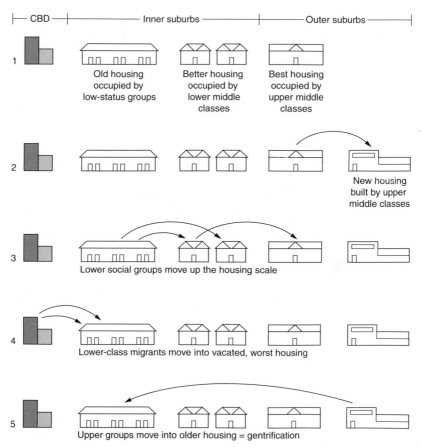

Figure 5.4 The process of filtering

4 Changing Residential Environments in Cities in MEDCs

Residential environments in MEDCs may undergo physical change in two main ways. They may experience **redevelopment** when old housing is demolished and replaced by new types of accommodation. They may however undergo **restoration** or **refurbishment** when the old housing stock is upgraded and made suitable for modern family needs. Both of these types of changes are widespread in Britain, and the way in which they are carried out today is a reflection of lessons learned in the past. Redevelopments are today much less radical than they were in the 1960s and 1970s, and housing developments today will involve the restoration of old buildings where possible. Gentrification, by its very nature, is a form of restoration and refurbishment.

Social and economic changes also take place within residential environments. Gentrification is one of the most obvious examples of this and is dealt with in some detail later on in this chapter. Another social change that certain city areas may experience is the domination by a particular age group. In Britain today, there is the relatively new phenomenon of **studentification**, which is also looked at below.

a) Residential redevelopments

The biggest phase of radical redevelopments in Britain took place in the three decades following the Second World War. Housing stock from the mid to late nineteenth century, particularly that located in inner city areas, was found to be very substandard. Whole areas had suffered war damage, but what was left of the nineteenth-century housing stock was often of a very poor quality. Just to take one indicator, the census of 1951 showed many inner city housing areas to have 30–50% of homes without inside lavatories.

Widespread redevelopments took place in all British inner city areas in the 1960s and 1970s and they were carried out by local authorities rather than the private sector. New building technologies such as prefabrication of wall sections enabled housing to be constructed cheaply and quickly, and also hailed the era of high-rise tower blocks. Town planners regarded large-scale public housing as the panacea to the shortage of homes and were influenced by both the Le Corbusier model in France and the perceived successes of house building in the Communist world.

Within two decades much of the new post-war housing was itself falling into disrepair and was creating a whole range of social problems. In 1968 a gas explosion in the Ronan Point tower block in London, which caused the collapse of one corner of the building, led to the questioning of the wisdom of creating high-rise developments. Prefabricated units appeared to encourage physical problems such as rising damp. Above all it was the social problems of isolation, high crime rates, drugs, vandalism and graffiti that came to stigmatise the post-war large-scale housing developments. With the selling off of a large proportion of local authority housing in the 1980s, the rise of the role of housing associations and the cost factor leading to local authorities opting for public–private partnerships, the whole nature of social housing has changed in Britain, and this is exemplified by the case study below.

CASE STUDY: THE HOLLY STREET REDEVELOPMENT, HACKNEY, LONDON

The Holly Street neighbourhood has seen two very contrasting redevelopments: one in the 1960–1970s and the other between 1992 and 2002. Originally an area of Victorian terraced houses of medium quality built in the 1880s, Holly Street and its adjoining streets were laid out on a typical grid pattern of that era. Throughout the twentieth century the houses became degraded and there was some damage to the area in the Second World War.

The whole area was redeveloped in the 1970s and the terraces were replaced by four tower blocks and a series of seven-storey 'snake blocks' which snaked across the neighbourhood totally destroying the old street pattern. By the late 1980s the new housing had become so degraded that the estate had the full range of problems associated with poor housing: social isolation, high crime rates, high rates of drug use, graffiti, vandalism, and services such as heating and lifts not working properly. The snake blocks proved to be worse than the tower blocks as they provided the ideal environment for burglary and mugging.

This most recent redevelopment was carried out between 1992 and 2002 in order to put right the mistakes of the 1960s and 1970s. It was a public–private initiative by Hackney Borough Council in conjunction with Levitt, the designers, Boris Construction and Laing Homes. The new neighbourhood has gone back to a similar grid pattern to that of the original Victorian estate. All of the snake blocks were demolished and all but one of the tower blocks, and that was re-clad and upgraded. The new terraced houses have front and back gardens, they are of different sizes to accommodate different types of family and have induced a completely different attitude from local people. During the redevelopment 3000 local authority rented homes were demolished and replaced by 800 housing association homes and 1200 houses that were for sale. Seventy-two of the units provide sheltered accommodation for the elderly. Traffic-free playground areas have been created for the children of the area. There are five housing associations that represent different interest groups, one of which is the North London Muslim housing association. The security of the area is enhanced by security guards and surveillance cameras.

The Holly Street estate is now held up as a model for how this type of development should be carried out. It has seen the replacement of a severe high-rise development with a greener medium-density housing environment. However, only 10% of the families living on the estate were there before redevelopment, and this has caused some problems in rehousing the rest.

b) The process of gentrification in MEDCs

The term **gentrification** was first recognised and coined by Ruth Glass working in London in the 1960s. She saw various working class districts of inner London being taken over by middle class families and as a result witnessed an upgrading of these neighbourhoods and their housing stock by the incoming new 'gentry'. As she put it:

> 'One by one, many of the working class quarters of London have been invaded by the middle class ... Once the process of "gentrification" starts in a district it goes on rapidly until all or most of the original working class occupiers are displaced and the whole social character of the district is changed.'

> Glass (1963)

Gentrification in London is not just restricted to districts of traditional working class housing, but also occurs in areas that had prosperous middle class housing when first built, but then became degraded in subsequent decades. Thus in London (and in other British cites), gentrification may affect a variety of housing types including:

- late Georgian and early Victorian middle class terraced houses that have become degraded and split up into multiple occupancy (e.g. bedsits)
- large late Victorian detached houses and 'villas' that have also become degraded and split up for multiple occupancy
- smaller late Victorian working class terraced houses, generally of the 'two-up, two-down' room layout type.

In the great financial boom of the 1980s a lot of high-earning young professionals were involved in the gentrification process. In London there was considerable change as the metropolitan area was undergoing one of its fastest periods of transformation from an industrial to a post-industrial city. This 'Thatcherite' (named after the then prime minister) decade saw considerable polarisation between the high earners and low earners within the capital. Younger people who made their money in the City and other areas such as the Docklands were known as **yuppies** (an abbreviation of 'young upwardly mobile professionals'). These people were particularly interested in living relatively close to the city centre and fired the gentrification process to such an extent that the areas into which they moved were labelled as being **yuppified**.

Within inner London, some of the most striking examples of gentrified neighbourhoods include Notting Hill, Battersea, Camden Town, Kentish Town, and various parts of Islington, including Canonbury and Barnsbury.

CASE STUDY: GENTRIFICATION IN BARNSBURY, ISLINGTON, LONDON

Lying between the Angel and Holloway, Barnsbury is a very distinctive neighbourhood of the London Borough of Islington.

Housing in Barnsbury, London

Most of its streets and squares were developed between the 1820s and the 1850s, when there was great demand for new housing for middle class people with jobs as clerks and middle management in the City of London, just a few kilometres to the south. The area started to go into decline in the 1920s when extended underground train lines enabled the middle classes to move further out into suburbia and have larger houses with gardens. From the 1920s until the 1960s, the general levels of the housing stock within the neighbourhood declined fairly rapidly. By the middle of the twentieth century, therefore, much of the area had become very run down and, after the Second World War, the borough council demolished and redeveloped some parts of Barnsbury, creating both high-rise and low-rise local authority housing; it also acquired whole streets and squares of older buildings and split them into flats for rental.

Gentrification started in some parts of Barnsbury from the late 1960s onwards, but progress was relatively slow. From the early twentieth century right through to the 1960s, certain parts of Barnsbury were almost 'no-go' areas for the general public as they were dominated by street gangs. Housing stock had become very run down and private land-owners split many of the fine old houses up into numerous bedsits with poor-quality services, renting them out to the highly mobile new migrant population. Once gentrification began, there was a reversal of the situation and bedsits and small flats were converted back into larger flats and whole houses. Jonathan Raban was living in the area when gentrification began and comments on the types of changes that took place in his book *Soft City*:

> 'From then on, estate agents' boards began to sprout from the basements, and builders and interior decorators swarmed round the square, carrying whole walls away from inside houses, pointing the brickwork, painting the fronts, taking long speculative lunch hours in the pub, while Nigel and Pamela, Jeremy and Nicola, made spot checks in their Renaults and Citroëns ... (In three years, a typical house on the terrace rose from £4000 to £26 000.)'

One of the reasons why Barnsbury became an attractive place to the gentrifiers was its early adoption of a traffic control scheme. As Lysett Green (1996) observed:

> 'In the early sixties, David Wagner, a resident and town planner, put forward a proposal to keep Barnsbury's "village" character by excluding major traffic from its centre. By 1970 it had become London's first major experiment in traffic management. It is now a maze of blocked off street entrances laced with sleeping policemen which serve to infuriate passing drivers and delight its inmates. You *have* to walk in Barnsbury and that cannot fail to make you happy.'

Since Islington Borough Council changed political complexion in 2001 from Labour to the Liberal Democrats, traffic-calming schemes have been even more vigorously pursued and restricted parking zones in the Barnsbury neighbourhood have been systematically extended.

Some of the main effects of the gentrification process in Barnsbury include:

- The closing and conversion of shops and pubs to residential accommodation; this reflects both the competition they have had from supermarkets and 'super-pubs' and the fact that residential land use can be more profitable than small-scale commercial land use; in one street alone (Cloudesley Road) all 23 of its shops (except one which survives as a woodcarver's workshop) were converted to residential use during the 1980s and 1990s. Furthermore since 1990, 11 'local' corner pubs have been closed and converted to flats, one has been converted into an evangelical church and two have become expensive restaurants; many of the remaining locals have become 'gastro-pubs' catering for the newer elements within the area.

- Between 1971 and 2001 the percentage of people in manual and semi-skilled occupations declined from 97% to just 36%, reflecting the influx of professional people to the area.
- Between 1971 and 2001 the percentage of owner–occupation of houses rose from just 23% to 62%, reflecting the housing demands of the newcomers.
- There has been a huge decrease by 40% of privately rented accommodation; of this half was bought up by the council and half bought by property companies or individuals and are now of owner–occupier status.
- By 2001 every ward of the Islington South and Finsbury constituency had fallen from Labour to the Liberal Democrats on the local borough council (including the wards which form part of Barnsbury) – even though Tony Blair was a local resident until 1997.

c) The 'studentification' of residential environments

In Britain in the late 1990s, a new form of gentrification was recognised and widely studied, that of **studentification**. This involves the development of large student communities concentrated within certain inner suburbs of university cities. These suburbs tend to be those close to both the university faculties and the leisure and entertainment facilities of the city centre. Typically the housing is relatively cheap – in such forms as Victorian terraces – and privately rented. Speculative land-owners may buy up whole rows of houses specifically for lucrative student rental. With the rapid expansion of universities in most British cities in the 1980s and 1990s, undergraduates now form a significant proportion of the population of places such as Brighton, Bristol, Leeds, Loughborough, Manchester and Nottingham. The phenomenon of large concentrations of students within certain city suburbs was recognised as early as the 1980s and various authors referred to them as 'student ghettos'. It was, in particular, the work of Darren Smith and Louise Holt of Brighton University in 2003 that gave a more detailed analysis of the studentification process; this also led to the phenomenon being given considerable press coverage.

The situation within Leeds has reached the point where the city authorities are taking measures to try to engineer the population structures of the areas of the city where the student population is most concentrated: Headingley, Burley and Hyde Park. Of these three areas, Headingley is the most studentified as the main halls of residence are located there as are all 27 private letting agencies which deal with the student market. Between 1997 and 2000 an estimated 1600 houses were converted to student use in Headingley and some 8500 families left the district as a consequence; during the same period, property prices rose by 50%, putting a lot of first-time buyers out of the market. Some of the knock-on effects of studentification in these areas include the proliferation of cheap alcohol outlets, the conversion of traditional pubs to theme pubs which may close in the summer months when the students are on holiday, the conversion of inner city factories into student accommodation, and school class sizes that have dropped to the extent of threatening the closure of schools.

The reactions to having large concentrations of students within inner city Leeds are mixed; on the one hand there are property owners who profit from the short-term lets, and on the other hand there are residents in Headingley and elsewhere who find it either a threat or a nuisance. The policy adopted by the city council is known as ASHORE (Area of Student Housing Restraint). This has involved the establishment of a zone of four square miles in the inner suburbs that is an exclusion zone for new student halls of residence and for the conversion of existing housing stock into flats for students.

5 Residential Variations in LEDC Cities

The much faster population growth rate in LEDCs has led to more intense social and environmental problems than are normally experienced in MEDC cities. In particular the social polarisation tends to be greater, and this is reflected in greater extremes of housing environment. Figure 5.5 illustrates many of the forces at work. The main mechanism behind urban change is the large number of incoming migrants from the poorer rural provinces. As was examined in Chapter 1, one of the most significant differences between MEDC cities and LEDC cities is in the location of their richer and poorer districts. In LEDC cities the poorest residential areas tend to be on the peripheries of cities – a refection of the fact that the new incoming migrants are generally the poorest element within the city.

The poorest housing in LEDC cities is found in the **shanty towns** or **squatter settlements**. These are the illegal settlements established by the poorer migrants to the cities when they cannot find or afford any other type of accommodation. Typically built of any available material – wood, corrugated iron, plastic, cardboard – these dwellings are known by a variety of names according to the part of the world in which they are found. Names for these 'spontaneous settlements' include *favelas* in Brazil, *pueblos jovenes* or *barriadas* in Spanish-speaking Latin America, *bidonvilles* in North Africa and *bustees* in northern India. Although concentrated on city edges, shanty towns may be located on any free land within an urban area.

It is not unusual for the poorest categories of housing to account for more than 50% of the population of an LEDC city. A survey

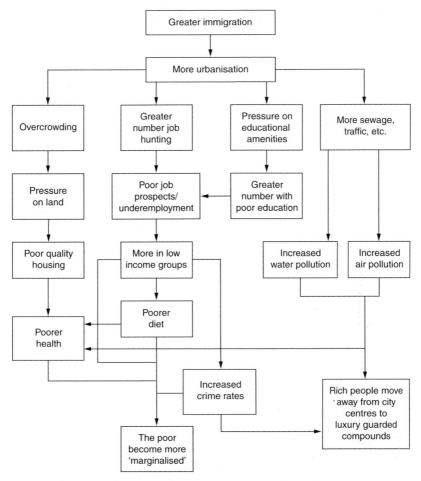

Figure 5.5 Dynamic changes coming from urbanisation in LEDCs

carried out in Cali, Colombia, categorised housing quality into six groups. At the bottom level are the newly built and poorest shanty towns in which 19% of the people live. At the next level are the shanty towns which have been in existence for longer, and which although they may have been slightly upgraded, are generally still one-roomed and still lack basic amenities; these account for a huge 40% of the population. At the next level up are the simple non-shanty homes of the urban poor, where a further 23% of Cali's population lives. Middle class homes of a basic type house are lived in by 7% of the population and larger middle class homes accommodate a further 8%. At the top of the social scale of housing are the luxury apartments and villas of the rich, which house just 3% of the population.

6 The Location of Shanty Towns within Cities in LEDCs

Just as shanty towns are built from any available material, they are located on any available plot of land within or around a city. The biggest concentration is found around the periphery where poor agricultural land and unproductive land such as deserts and mountains may be found. Rich farmland is less likely to be turned over to squatter settlements. Shanty towns are also strung out along lines of communication into the big cities – particularly the roads leading in from the poorer provinces from which the migrants have come. Within cities, different types of wasteland, such as that along railway lines, close to factories or near warehouses and transport depots, provide locations ripe for shanty developments. Places with poor drainage, yet which have water supplies available, such as riversides, canalsides and marshes, also attract squatter settlements. Even close to or within the CBD, temporarily vacated sites where new developments are about to take place also become attractive spots for new migrants to settle.

CASE STUDY: THE LOCATION OF SHANTY TOWNS IN LIMA, PERU

An estimated 50% of Lima's 6 million people is to be found living in shanty towns. Figure 5.6 shows where the main *barriadas* are located. With an annual average rainfall of just 27 mm, it has a hyper-arid climate and is surrounded to the north and south by the desert coastal plain of the Pacific; to the east are the arid foothills of the Andes. Through the city runs the Rio Rimac, a river that is very irregular in its flow pattern. Not surprisingly, water supply is the biggest problem faced by the shanty towns of Lima, and public health issues, particularly diseases associated with poor sanitation and polluted water, pose a great problem among the children of the *barriadas*.

The map shows that there are large concentrations of shanty towns in the industrial zones lying between the city centre and the main road down to the port of Callao; Callao itself is a relatively poor part of Greater Lima. Large concentrations are also found along the desert coastal plains to the north, around Comas and to the south, around Chorrillas. Here the extensive *barriadas* spread over land of little agricultural value and are strung out along major highways. The biggest of all Lima's shanty

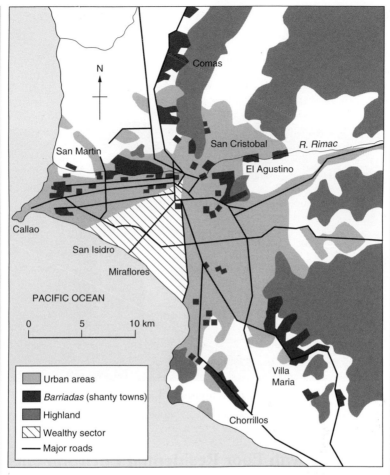

Figure 5.6 The location of shanty towns in Lima, Peru

towns is Villa Maria, which houses an estimated 700 000 people and spreads from the desert up three mountain valleys. Land along the Rio Rimac also has several large shanty towns; there are even small *barriadas* along the river very close to the historic city centre. The most noticeable absence of shanty towns is to be found in the wealthy sector of Lima, which leads from the city centre to the seaside suburbs of Miraflores and San Isidro.

Shanty town houses in Lima, Peru

7 Dealing with Poor Residential Environments in LEDCs

Shanty towns change through time and they generally undergo improvements in the quality of life they offer their inhabitants as time progresses, for the variety of reasons discussed below. The achievement of these changes very much depends on the wealth of the country or city in question, and greater progress has been made in the last few decades in Southeast Asia and Latin America than in sub-Saharan Africa.

Four theoretical stages of shanty town development can be recognised:

- the **'invasion' stage** when new migrants arrive and establish their simple shelters as their first urban home
  the **consolidation stage** which occurs within 1–10 years of their establishment; although the actual buildings may still remain

illegal, without basic amenities and in a very poor state, their are minor improvements in their inhabitants' quality of life
* the **upgrading stage** which occurs between 10 and 20 years after their establishment; during this time self-help improvements and the laying on of basic amenities such as water and electricity greatly improve the quality of life of the inhabitants
* the **assimilation stage** which generally occurs after 20 years, when the shanty town fabric has been so improved that it becomes integrated as a normal and legal part of the city.

Just as shanty towns may not pass through this full sequence for economic reasons, there may be restraints brought about by physical geography. By their very nature, shanty town homes are more vulnerable to natural hazards than more substantially built houses. In 1997, for example, Hurricane Mitch made tens of thousands of poor people in Tegucigulpa, the capital of Honduras, homeless. Torrential rain swept away the flimsy shanty town structures on a series of deforested hillsides overlooking the river, close to the city centre.

The approaches to shanty town development vary from country to country and city to city, and each place is unique, its approach reflecting local political and economic conditions. For the sake of convenience the treatment of shanty towns can be put into the following categories:

* The **'do-little-or-nothing' approach**, which is the product of powerlessness to cope with the sheer size of the problem and the volume of newcomers setting up spontaneous homes. In many sub-Saharan African states housing problems have been made worse by conflicts and the influx of refugees. For example, 96% of urban dwellers in Sierra Leone live in slum conditions. Mogadishu, the capital of Somalia, is a barely functioning city since the civil war of the 1990s (see the Case Study on pages 126–7).
* The **slum clearance and do-nothing approach** is, while a cheap option, the most short-sighted one as it creates the conditions for civil unrest. When an earthquake destroyed much of Managua, Nicaragua, in 1972 a great deal of the poorer housing was flattened. Money for reconstruction was siphoned off by the ruling family and hundreds of thousands of people were displaced. The large body of the dispossessed poor eventually led to the Sandinista Communist takeover of the country and the problems of housing were partially solved. The most ruthless example of this approach was in mid-2005 when President Mugabe of Zimbabwe authorised the destruction of the squatter homes of some 200000 people through a politically motivated policy known as *Gujuranhuni* ('Cleansing').
* The **clearance and new building approach** is an expensive but effective option. Wealthier cities and countries in parts of the world such as Southeast Asia have been effective in using this

approach. Singapore's housing policy (considered in a Case Study below) has in particular been held up as a model to the rest of the world. The danger with this approach is the replacing of slums with cheap housing that quickly disintegrates into new forms of slums, as has been the case with some of the high-rise estates in cities such as Mumbai.

- The **New Town approach**, which is similar to that above, but takes pressure off the city by building new housing areas at some distance from the city itself. The Case Study of Cairo below provides a good example of this approach. Hong Kong and other Chinese cities have also put this into practice.
- The **improvement of existing housing approach**, which comes from two main sources, the intervention of the local authority and the shanty town dwellers themselves through **self-help**.

There are two main types of improvement approach:

- **upgrading**, where the role of the local authority is dominant; one of the earliest and a relatively successful example of this has been the **Kampung Improvement Programme** in Jakarta in Indonesia. The city government has laid on basic amenities and helped to upgrade the housing itself
- **site and service**, where the local government lays on some amenities, but encourages the slum dwellers to upgrade their buildings often with grants or charity aid. This has proved very successful in countries as diverse as Botswana, Tunisia and Pakistan.

CASE STUDY: SINGAPORE'S PUBLIC HOUSING SOLUTION

As such a successful NIC, Singapore cannot in any way be regarded as an LEDC today, yet when it embarked on its public housing policy as an answer to the problems of substandard residential areas 50 years ago, it was very much a developing country. Singapore has now experienced almost 50 years of housing developments, during which the government has replaced both traditional *kampongs* (Malay-style villages in the countryside) and substandard urban housing. With the setting up of the Housing Development Board (HDB) in 1959, the government embarked on a series of five-year plans. In the 1960s over 100 000 (mainly high-rise) units were constructed and a total of 23% of the population was rehoused. These early HDB units were basic with either communal or semi-communal toilets, and most were built 6–8 km from the city centre to enable easy transport access.

In the 1970s when over 200 000 units were completed, the basic standards were much higher and all homes had inside bathrooms and kitchens. Some of the estates were effectively new towns and their development coincided with the growth of the public transport network (at this stage roads and buses). The style of building remained very utilitarian, but an increasing proportion of the people were being encouraged to buy their homes. The 1980s saw the period of diversification, with new styles of architecture making each estate more individual, and there was also a much greater sense of building integrated communities with a wide range of facilities including schools, sporting amenities, shops and community centres. By the end of the 1980s, 86% of Singaporeans lived in HDB housing.

The 1990s and the twenty-first century have seen the concentration of the HDB on the upgrading of existing housing to suit the demands of the new, wealthier and better-educated generation. As the richer middle classes now opt for luxury condominium housing built by the private sector, over 70% of the population still relies on HDB housing.

CASE STUDY: CAIRO'S NEW TOWN SOLUTION

Cairo, Africa's largest city, has an estimated population of somewhere between 8 and 12 million people. The vast numbers of people living in slum accommodation and shanty settlements in areas of poor environmental quality, such as the heavily polluted industrial area of Fustat, the cemetery area known as the 'City of the Dead' and the various city rubbish dumps, heighten an already severe housing shortage. In 1992 an earthquake hit Cairo and thousands of poorly constructed mud-brick buildings collapsed, leading to even higher levels of homelessness.

As early as 1969, the authorities planned to rehouse a large number of Cairenes in New Towns at some distance from the city (see Figure 5.7). Four New Towns were founded in the 1970s and were planned to be part of a total of an eventual 14. The first wave of New Towns lies around 40 km from Cairo and the population is supposed to reach an eventual 250 000–500 000. At a greater distance from Cairo was the next generation of New Towns which are to have eventual populations of between 500 000 and 1 million. The 10th of Ramadan is one of this second wave of New Towns and has proved to be a success story. Fifty-five kilometres from Cairo and established in 1977, it had by the late 1990s attracted over 800 industrial plants creating some 60 000 new jobs for skilled workers on salaries 30% higher than those in Cairo.

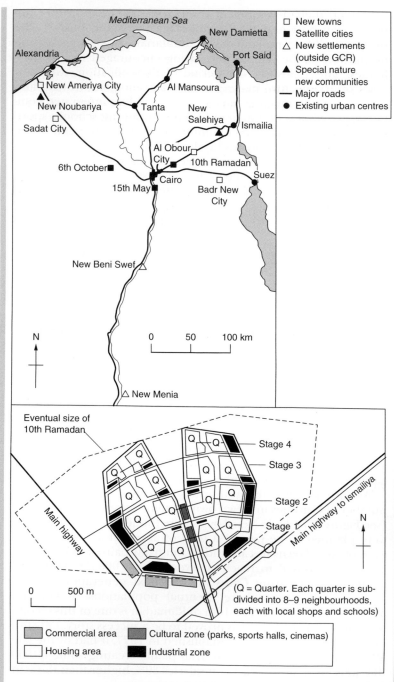

Figure 5.7 Cairo's new towns

The plan of 10th of Ramadan is cellular, allowing for easy and logical expansion. Along with the success there are however problems; water supply is a key issue in this desert location. The housing is of a much better standard than much of that in Cairo, but a lot of it has remained empty and has been subject to property speculation. Also it could be claimed that this and other new towns have not done as much as expected to solve the Cairo housing problem as it has been mainly the skilled workers and not the poor who have settled there.

Questions

1. Using Figure 5.1, explain why the processes at work in residential areas in cities in MEDCs create such widely diverse social areas.
2. Explain the conditions and processes that give rise to **gentrification** in MEDC cities.
3. With reference to specific examples, examine the conditions under which so much **urban redevelopment** has been necessary in the UK in the last two decades.
4. Account for the location of shanty towns in cities in LEDCs. Assess the different ways in which the problems of these squatter settlements have been dealt with in different places.

6 Transport and Accessibility in Urban Areas

'Unreal City,
Under the brown fog of a winter dawn,
A crowd flowed over London Bridge, so many,
I had not thought death had undone so many,
Sighs, short and infrequent, were exhaled,
And each man fixed his eyes before his feet.
Flowed up the hill and down King William Street, ...'

TS Eliot *The Wasteland*

Commuting is probably the greatest curse of the contemporary city. With cities of ever greater proportions and populations, more people are on the move and they are travelling over greater distances. The consequences of this are often counterproductive, rendering slower journey times, phenomena such as gridlock and a deterioration in the urban environment through pollution. Not only does poor traffic circulation cause physical damage to the city, but it also causes economic damage as, to use the old adage, 'time is money' in the business world.

In pre-industrial Europe and in other parts of the world prior to 1900, most cities were small enough for people to make their journeys on foot. The Industrial Revolution, however, led to such a rapid expansion of urban settlements that new forms of transportation were necessary, and these solutions were found in the new technologies developed out of that revolution. Trains, trams, buses and cars have all revolutionised transport and urban accessibility in the past 150 years.

1 The Development of Transport Systems and Urban Growth

Figure 6.1 shows the relationships between various stages in urban development and the nature of the transport technology and transport system. The pre-industrial city was typically compact with traffic

	Urban functions	Transport technology	Transportation system	Urban form
Stage 1: Pre-industrial	Defence, marketing, political-symbolic, craft industry	Pedestrian, draught animal	Route convergence, radial	Compact
Stage 2: Early industrial	Basic industries, secondary, manufacturing	Electric tram, streetcar, public transport	Radial improvements, incremental additions	High-density suburbanisation, stellate form
Stage 3: Industrial	Broadening industry, tertiary service expansion	Motor bus, public transport, early cars	Additional radials, initiation of 'ring' roads (incomplete)	Lower-density suburbanisation, industrial decentralisation
Stage 4: Post-industrial	Additional of quaternary activities	Towards universal car ownership	Integrated radial and circumferential road network	Low-density suburbanisation, widespread functional decentralisation

Figure 6.1 The relationship between transportation and urban form in Western cities. (*Source:* D Herbert and C Thomas (1997) *Cities in Space; City as Place* London: David Fulton)

dominated by pedestrian and draught animal movements. To these modes of transport one might add both bicycles and small motorcycles. Already the embryonic main road pattern was a radial one.

Each successive wave of transport technology led to a strongly radial network of roads, railways and tramways. At the same time each wave encouraged the growth of suburbs and an ever lower building density at the city edge, as this moved further and further out. The cities created by the Industrial Revolution grew rapidly because of the development of various means of mass mechanised transport, especially railways and trams, that enabled at least the wealthier city workers to commute over greater distances.

In the current post-industrial phase of growth in MEDCs, the city is dominated by almost universal car ownership, a very complex network of roads and the highly decentralised functions associated with such phenomena as the Edge City (which will be dealt with in more detail in the next chapter). Los Angeles provides the most striking example of the transport systems and road networks associated with the post-industrial city. With its complex network of urban freeways (motorways), the very fabric of the city is cut up into a series of often unrelated blocks; this was discussed in Chapter 1 in the section on Los Angeles and the Dear and Flusty model of the post-modern city (see page 21). Even in such a car-dominated city, however, a considerable amount of traffic will be taken off the roads when the efficient multi-line MetroRail network, which first started operation in 1990, reaches its full 220 km of track sometime in the twenty-first century, when it will link many of the most important nodal points of Greater Los Angeles.

2 Traffic Congestion and Possible Solutions

Urban road networks have often developed over a long period of time, and each successive layer within this pattern reflects a different stage in transport technology. In the age of the motor car, traffic congestion in city centres, the volumes of rush-hour traffic and the increasingly complex patterns of movement require often radical solutions.

The two main ways of dealing with road traffic congestion are the imposition of restrictions on private transport and the improvement of public transport alternatives. This section will look at the former approach and the next section will consider the latter. Pedestrianisation of certain important squares and streets in central areas of cities has been practised for decades in many European cities. Although this form of traffic segregation has generally been successful, it affects rather limited sections of CBDs, and more draconian measures need to be introduced to solve congestion problems.

CASE STUDIES: LONDON AND SINGAPORE

Traffic congestion is a problem in nearly every city as a result of the growth of car ownership and the expansion of surface forms of public transport such as buses, trams and light rail schemes. The fact that most cities are dominated by radial road networks merely contributes to congestion. Where most of the commercial functions remain in the city centre, these radial roads focus traffic on the CBD, which is where congestion measures need to be taken. Each city has a different approach to solving these problems and the two examples of London and Singapore provide excellent contrasts in approach.

a) London

London's traffic experienced an unprecedented growth in the final three decades of the twentieth century. Although traffic jams are nothing new – at the turn of the twentieth century horse-drawn omnibuses and carriages regularly blocked the West End and the City of London during the daily rush hours – it was the growth of motorcar ownership between the 1950s and the 1990s that brought matters to crisis point. By the turn of the twenty-first century, motor traffic fumes had replaced domestic and industrial chimneys as the main source of air pollution, and the average traffic speed in central London had been reduced to just 17 km/hour at peak times.

When the Greater London Authority (GLA) was established, Ken Livingstone used **congestion charging** as one of his main policy platforms in the mayoral elections. Duly elected, he put forward his proposals in February 2002 and they were put in force a year later. Private traffic is charged a fee to enter the central area of London (the City, the West End and some other parts of Westminster) between 7.00 am and 6.30 pm on Mondays to Fridays. There are various means of payment in advance, but payment after 10.30 pm on the day of entry incurs a penalty fee. All of the traffic entering the central zone is electronically checked by cameras on approach roads.

The congestion charge has proved a great success as traffic circulation speed has risen to around 30 km/hour at peak times, there has been a 12% reduction in traffic pollution and there has been a 40% increase in the number of passengers on buses, concurrent with a 24% drop in waiting times at bus stops. On the negative side, there have been problems with extra traffic and parking congestion in the areas just outside the central zone. Access to shops within the central zone has also been problematic in some cases. The London Underground has, unlike the buses, seen a 6% decrease in passenger use, but this has been more to do with its high fares, engineering works and safety issues.

The GLA underestimated the success of the congestion charge and has therefore not received as much income from it as originally expected. Cynics suggest that this is the main reason why the GLA is considering extending the charge zone into mainly residential areas in Kensington and Chelsea and at the same time raising it by 40%.

b) Singapore

Singapore has long recognised the problems posed by mass car ownership on what is a small island (582.8 km²) with 3.6 million people, and has taken remedial action. At present there are some 700 000 motor vehicles in Singapore of which around 500 000 are private cars. Roads take up 12% of the island's land and the number of vehicle trips is growing at a rate of about 10% per year.

Since the 1970s, the government has controlled car ownership by a number of measures, including a 45% import duty and a whole range of registration fees, making car purchase and ownership expensive. In 1990, a vehicle quota system was introduced, making it yet more difficult to acquire a car.

In order to control traffic in the more congested central areas, the government introduced an area licensing scheme in 1974. Vehicles had to buy a licence to enter the CBD between peak hours on working days. In 1997 this scheme, which had worked successfully, but involved a lot of paperwork, was replaced by the ERP (Electronic Road Pricing) scheme. Cashcards are fixed to vehicles and are automatically read as they pass under a gantry over the road. The CBD has an entry charge between 7.30 am and 7.00 pm, and expressways have a toll between 7.30 and 9.30 am. The charge depends on the size and type of vehicle; for example taxis are charged less then private cars to encourage people to use them as an alternative. The effects of the ERP scheme were immediate, leading to a reduction of 25 000 cars at peak times and an increase in average vehicle speeds of 22%. It also encouraged car pooling and a consequent decline in solo drivers.

3 Public Transport Networks

Whereas the limiting of private transport in cities by banning it on certain days or in certain areas, or by congestion changes and road pricing can be regarded as negative approaches, the improvement of public transport can be regarded as a positive approach to congestion. The main problems of encouraging more people onto public transport include cost, safety, reliability, cleanliness and how comprehensive is the area covered by the networks. It is a question of dealing with both the experiences and the perceptions of the general public.

Public transport in cities falls into three main categories:

- metros: trains running on dedicated track, either above or below ground; although the most expensive and time consuming to create, they are the best solution to congestion
- trams and light rail schemes: these tend to have a mixture of dedicated track and road space and are less expensive than metros
- buses: these are the cheapest solution, but can both be the cause of congestion and be affected by it.

The three following case studies look at the relative success of these three modes of public transport.

CASE STUDY: THE SINGAPORE METRO

The route network of the Singapore Metro has evolved rapidly over the past two decades and can be regarded as a model for metro provision in other Southeast Asian cities. Until the 1960s a metro system would have been seen as unnecessary, but with the expansion of public housing and New Towns throughout the island of Singapore, a network was planned. In 1967 a four-year study was carried out by the Singapore government in conjunc-

The Singapore MRT

tion with the UN Development Programme. The first sections of the SMRT (Singapore Mass Rapid Transit) opened in 1987 linking some of the most densely populated parts of the south of the island with the CBD. As the housing schemes spread throughout Singapore, the network grew. One of the lines was extended in 1996 to form a loop going to Woodlands in the very north of the island. In 2003 the new North East Line was added to the network, linking the harbour and CBD to some of the new suburbs on the east coast. In the same year Singapore's Changi Airport was also linked up to the network.

The MRT is built to high specifications. The trains and underground stations are air-conditioned, which is important in the equatorial climate. The underground stations pioneered glass safety doors on platforms, of the type now used on the Jubilee Line extension in London. Suburban lines are overground, but elevated in order to save space on this densely populated island. The MRT is also graffiti free and extremely clean, to the extent that durians, a smelly tropical fruit, are not allowed on trains or stations. Signalling and ticketing are highly computerised; electronic ticketing of the sort used on the London Underground was introduced two years before the 'Oystercard' had its trial run in London.

Between 2000 and 2004 two LRT (Light Rapid Transit) lines were added to the network. These are two-car elevated monorail metros that effectively wind in and out of housing estates in the Bukit Panjang and Sengkang areas, and have both greatly reduced the local dependency on buses. A new Circle Line of the MRT is now under construction and this will provide a loop around the greater central area of Singapore, as well as increasing connectivity between existing parts of the MRT. At present the network is run by two private companies one of which runs a large part of the public bus network. The two companies work in harmony rather than as competitors and give Singapore an exemplary integrated public transport system.

CASE STUDY: LIGHT RAIL AND TRAMS IN BRITISH CITIES

When Tony Blair's Labour government came to power in 1997, many political pundits suggested that transport would be its 'Achilles' heel'. After several years in power many aspects of public transport remained in a chaotic state. While other European countries, such as Germany, France, Italy and Spain, had invested huge amounts of money in their high-speed inter-city trains, Britain was just catching up with repairs on its ailing privatised long-distance lines. At the same time, the USA along with European countries such as Germany, France and the Netherlands had also been investing heavily in light rail and tram systems in their urban areas.

Many of Britain's large provincial cities had trams until the 1950s and 1960s, when the tracks were pulled up in favour of both the private car and the public bus. Similarly, as a reaction to the urban traffic congestion of the 1990s, many of the same large provincial cities drew up plans for light rail and tram systems, but by 2004 only those in Manchester, Birmingham, Newcastle, Sheffield and the south London 'Tramlink' based in Croydon were in operation. In the same year, many months behind schedule, the first stretch of Nottingham's new tram network was opened. Future prospects of tram or light rail systems in cities such as Bristol, Portsmouth, Liverpool, Hull and Cardiff have been put in doubt as a result of the government spending review in mid-2004. It is above all the problems of funding and the relationship between central and local government that have given the British tram networks an insecure future.

By the spring of 2005 the British government was backing a 'Streetcar' bus with a capacity of 130 passengers in preference to trams. These articulated buses, which look like trams, are a much cheaper alternative and they would cost just one-tenth of trams because they would run on dedicated road lanes rather than metal tracks. The Streetcar is to be piloted in York and many other British cities are likely to adopt it, including Sheffield, Leeds, Swansea, Reading, Bath and Glasgow.

In many continental European cities, where there has been a long tradition of tram use, and this was not interrupted by decades in which buses and cars replaced trams, there has also been little of the small-minded haggling over funds which has held back progress in the UK. In Britain the main determinant

Figure 6.2 Manchester's MetroLink network. (Copyright and reproduced by kind permission of GMPTE.)

of the nature of public transport has become profit making rather than the provision of a public service. The privatisation of public transport and the deregulation of buses in the mid-1990s added a great deal to the confusion and insecurity of funding for the nascent tram networks in British cities.

The Manchester Metrolink can be held up as a great success story that should be of encouragement to other British cities (see Figure 6.2). Manchester has two major railway termini, Piccadilly Station on the south-eastern edge of the centre and Victoria Station on the northern edge. There had long been plans to link these stations by a tunnel, but the costs of such a project, together with the nature of Manchester's underlying clay strata, meant that they were never realised. By the late 1980s the equipment on an electrified overland railway from Victoria to Bury was needing replacement, so this line became the basis of a new metro system. Costs and delays once again affected decisions and an overland tram system was adopted rather than an underground railway. In 1992 the first stretch of Metrolink was opened from Bury to Altringham via the city centre, with a spur to Manchester Piccadilly. The Phase Two extension, from Cornbrook to Eccles via the Salford Quays development, opened in 1999.

The Third Phase of the Metrolink would link up to the existing network Manchester Airport, and the town centres of Rochdale and Oldham. In July 2004, however, central government did not give the scheme financial approval, just as projects were shelved in other cities. Yet again public transport was being subjected to government 'stop-go' short-term policies. Whether or not Metrolink Phase Three will ever go ahead remains to be seen.

CASE STUDY: CURITIBA, BRAZIL: A MODEL FOR LEDC CITIES

Curitiba, the capital of Brazil's Paraná State, has a population of 2.3 million, but is growing at a rate of around 5.7% per year. Unlike many other cities in LEDCs, Curitiba has planned its public transport system in great detail over a long period of time. Ever since the 1940s the city has worked from a series of master plans, and as early as the 1970s structural avenues were established as a part of land-use planning. These were designed to encourage linear growth of both commercial and residential land use along radial routes that would be provided with efficient public transport. Thus, instead of putting in new transport lines to cope with an expanding metropolitan area, as happens in so many places, the planning authorities of Curitiba used their public transport strategy to determine how the city would grow.

When planning the radial routes, the local authority of Curitiba established some for incoming traffic, some for outgoing traffic and others, known as *canaletas*, for buses only. Where buses share roads with cars, buses have priority signalling. With over 2000 buses, despite having the second highest car ownership in Brazil, 70% of commuters do not use their cars and as a consequence Curitiba has only 30% of the air pollution of other Brazilian cities. There are three different types of buses used in the city and a whole range of different routes – such as express, neighbourhood, educational, night and park-and-ride routes – operated by the ITN (Integrated Transport Network) authority. For efficiency, the bus stops, which resemble stations contained within tubes, are elevated and sheltered, and fares are paid at a turnstyle before boarding. As the network only costs the equivalent of £150 000 per kilometre to establish (about one-fiftieth of what a metro line would cost), fares can be kept low and this encourages more people to use it.

The success of Curitiba's bus-based rapid transit system has led to its being emulated elsewhere; although in many ways it is the model of the cheap alternative to trams and underground railways for LEDC cities, it has been adopted in some MEDC cities. Elsewhere in Latin America, metro bus systems have been adopted in Quito, Ecuador, and in Porto Alegre and São Paulo (both in Brazil) – in the latter case to supplement the existing metrotrain network. In North America, the Canadians have been particularly interested in the bus systems that have been adopted in Quebec City, Vancouver and Montréal – in the latter example adding to the underground metro already in existence. Also, many US cities, particularly those with budget problems, are now looking into this cheap alternative to a metro network.

4 Pedestrianisation in City Centres

Although there is some evidence of pedestrianised areas in ancient cities, for example, there are still stone barriers intact in the ruins of Pompeii which prevented Roman charioteers from entering the city's forum (its main public square), in general there was probably very little segregation of people from traffic in pre-industrial cities. Pre-industrial traffic forms do not preclude traffic jams, however. In the western Chinese city of Kashgar the police have to be out at Sunday morning markets to direct the horse-drawn vehicles in order to prevent absolute chaos.

In the modern world it is motorised transport that is being segregated from pedestrians and this is done for a variety of reasons in various cities of the world. Within MEDCs, historic pre-industrial urban cores particularly lend themselves to pedestrianisation. In British city centres such as those of Lincoln, Canterbury and Winchester there is some degree of pedestrianisation of historic streets. Most of the New Towns (see Chapter 2) that were developed in Britain after the Second World War had by their very design small pedestrianised commercial districts in their centres.

However, in the past decade or so, it has been the larger industrial cities that have undergone more dramatic pedestrianisation; this has been due to three main changes:

- the growth of retail concentrations along the main high streets and adjoining streets and the consequent high pedestrian densities
- the opening of large city centre shopping malls which by their very nature are pedestrian only and the way in which they are linked to high streets
- the development of specific new cultural, retail and entertainment facilities in disused docks, industrial areas and along canal sides, which have been designed to be pedestrianised.

Cities such as Leeds, Liverpool, Bristol, Birmingham and Nottingham all have examples of these types of developments. London has been slow to create pedestrianised precincts and the city has a few

Pedestrianisation in Skopje, Macedonia

successful but unlinked examples including the Covent Garden *piazza* and adjoining streets and the walkways along the Embankments of the Thames. The bold scheme to pedestrianise Trafalgar Square, Whitehall and Parliament Square was partially realised in 2004.

In general, continental European cities have been more proactive in pedestrianisation and the preservation of historic city centres. In Italy, Bologna is held up as the model of traffic control and pedestrianisation of historic streets because of the foresight of its city leaders in the 1970s and 1980s. Elsewhere cities such as Milan, Turin, Florence and Rome also have very successfully segregated areas.

Many cities in the transition economies of Eastern Europe also have very successful pedestrianisation schemes. They have had the advantage of relatively low levels of private car ownership until very recently. Debrecen, the second city of Hungary, and Kosice, the second city of Slovakia, have very effectively pedestrianised their main, broad cobbled 'high streets' which are lined with shops, cafés, restaurants and historic buildings. In the case of Debrecen, a tram remains the only form of transport along this main street, whereas in Kosice, even the tram has been removed. In Skopje, the capital of Macedonia, there is a highly effective pedestrianised system of streets which link the old pre-industrial Turkish quarter with its mosques, shops and restaurants via a steeply arched Ottoman footbridge to the city's main traffic-free Makedonija Square (which under Communism was used for military parades). This in turn is linked to a pedestrianised way along the River Vadar lined with cafés and bars and to a new multi-storey, indoor, ultra-modern shopping complex.

In many US cities, high reliance on private transport as well as grid patterns of streets have mitigated against large-scale pedestrianisation. As in Britain, the main stimulus for pedestrianisation can be found in places with important historic buildings (e.g. the colonial streets of Philadelphia, Pennsylvania); in places with a high concentration of shops, restaurants and entertainment facilities (e.g. the riverside walks at San Antonio, Texas); and where new developments of former port and industrial spaces have enabled pedestrianisation from the outset (e.g. Seaport South, New York City).

In LEDC cities, the situation is very mixed. There are the advantages of low private car ownership, but these are often outweighed by high densities of pedestrians, public transport and freight vehicles in city centres. India provides an interesting example of a country where cities have important historical buildings in need of restoration, high traffic densities causing congestion and pollution, and at the same time, rapidly expanding economies. In 1999, the urban authorities of Hyderabad in Andhara Pradesh embarked on a scheme to restore whole historic sectors within the central area of the city and at the same time to introduce pedestrianisation to many of these areas. It may well prove to be a model to be copied by other Indian cities, and indeed LEDC cities elsewhere.

Questions

1. Outline some of the main ways of dealing with traffic congestion in urban areas.
2. Why are public transport systems likely to be more successful in some cities than others?
3. Examine the way in which trams and light rail systems have been introduced into cities within the UK. Assess the success that such systems have had on those cities into which they have been introduced.
4. How can LEDC cities develop mass rapid transit schemes that cost less than metros in MEDC cities?

7 Life on the Edge

'A suburb must always be parasitic on a town or city. Even when it acquires an independent administration, it is never a financial centre, or a centre of power. Suburbs were rarely meant to be agriculturally or industrially productive.'

Joseph Rykwert *The Seduction of Place*

Suburbs rarely get a good press. In contrast to the excitement of what goes on in the city centre and the dynamic changes or shocking decay of the inner city, suburbs are, rightly or wrongly, perceived as dull, safe and pedestrian. A British television series of the 1990s was entitled '*Heaven, Hell and Suburbia*', and this left the viewers to consider the suburbs as neither one thing nor the other, but as a kind of in-between 'limboland'. In the USA the suburbs of large cities are often referred to by urban geographers as the **middle landscape**, for this very reason.

One of the most obvious and frequent criticisms of suburbs is the monotony of their housing types and the anonymity of their streets. This is equally true of the low-density suburbs of Britain, North America and Australasia as it is of the high-rise suburbs of the cities of continental Europe and Southeast Asia.

This perception of suburban housing is nothing new; in the mid-nineteenth century, Benjamin Disraeli in his political novel *Tancred*, complained about the similarity of the housing in the inner suburbs of London – now very fashionable and expensive terraces and squares, such as those near Paddington, St Pancras, Baker Street and Marylebone. He observed:

> 'Though London is vast, it is very monotonous. All of those new districts that have sprung up within the last half-century, the creatures of our commercial and colonial wealth, it is impossible to conceive anything more tame, more incipid, more uniform.'

Part of the reason for the uniformity was that Acts of Parliament had proscribed the size, style and building materials of these new developments. Yet not everyone has been a critic of the suburbs. A century after Disraeli, the poet John Betjeman frequently wrote with affection about the suburbs of Middlesex that had grown as a result of the overland extension of the Metropolitan Line of the London Underground in the 1930s, and found a certain magic in the place names, which he capitalised:

> 'Smoothly from HARROW, passing PRESTON ROAD,
> They saw the last green fields and misty sky,
> At NEASDEN watched a workmen's train unload,
> And with the morning villas sliding by ...'

This so-called 'Metroland' admittedly had much greater variations in building style than are found in the inner suburbs. This is due to the way in which they expanded piecemeal, using a variety of different planners and architects.

Now in the twenty-first century, new forms of suburbs and suburbanised countryside are developing at ever greater distances from city centres and even the cities themselves. These distant suburban developments are most noticeable in the USA and, as would be expected, they come in for some harsh criticism.

1 Suburban Growth, Counterurbanisation and Suburbanisation

The growth of cities is at its greatest when industrialisation takes place and rural–urban migration is at its height. In MEDCs the period from the early nineteenth century through to the mid-twentieth century saw the highest rates of urbanisation and consequent urban growth. In LEDCs, many cities are experiencing their highest growth rates to date.

Urban growth has invariably resulted in the extension of cities into their surrounding countryside and the building of housing and other types of infrastructure on what was before farmland or open spaces; this is what was being lamented as early as 1814 by Wordsworth, as shown in his poem quoted at the beginning of this book. It was the greater mobility created by the technologies of rail and road transport which stimulated suburban growth in Europe and North America.

The railways had the first major impact on suburban development from the mid-nineteenth century, followed by that of the motor car from around 1920 onwards.

Suburban growth has tended to be irregular, at first being concentrated along the main radial routes, thereby giving commuters a high degree of accessibility to city centres. This form of growth is therefore known as **radial growth** and **ribbon development**. Uneven development leaves wedges of land not built on within the suburban fabric, and this has been in the twentieth and twenty-first centuries a constant problem for town planners and city authorities. In some cities such as Rome the green wedges have become an important part of the protected land and green belt policy (particularly as in Rome some of these wedges have important archaeological sites on them). In most cities, however, this type of land is seen as an opportunity to create new housing by **infill**. As the outer suburbs of cities in MEDCs generally have the richest and most vocal members of society, land-use conflicts arise when new infill housing is planned; the **Nimby** (Not In My Back Yard) attitudes are strongly represented in the outer suburbs. The most extreme form of this attitude has been dubbed in the USA **Banana** (Build Absolutely Nothing Anywhere Near Anything). However, occasionally local residents may welcome a new building complex in their neighbourhood, for example, if it brings in high-status professional jobs and has a positive effect upon property values; in this case the situation is a **Yimby** (Yes! In My Back Yard) one.

Conditions are generally rather different on the edges of LEDC cities. As was seen earlier in this book, the land-use and residential arrangements in LEDC cities are generally quite different from those in MEDC cities. With the outer edges of cities frequently dominated by illegal squatter settlements and the poor legal housing of new migrants, suburban growth is often completely haphazard, and not necessarily a reflection of arterial routeways; infill takes place as new migrants occupy any available plots of unused land. Green belt and land protection policies are generally an unattainable luxury, on grounds of both cost and lack of ability to control the growth of poor and illegal housing. What is beginning to change this situation is the spread of suburbs of the rich, often gated communities into the outer urban fringe. This causes localised conflicts between rich and poor and can influence local authorities to take action – generally in favour of the rich. Nimbyism is not just the preserve of MEDCs.

In the last few decades in MEDCs **counterurbanisation** has been the process most responsible for population growth in both suburban areas and zones beyond the urban edge.

Counterurbanisation has taken place as a result of three main factors:

• the pressure on land in city centres and the outward expansion of CBDs into inner suburbs

- the decline and decay of the housing stock and service provision in inner city areas
- the quest of upwardly mobile groups for a healthier lifestyle in the outer suburbs and beyond.

Although many books suggest that counterurbanisation started in the 1970s, this was in reality when it was first noticed and widely written about. In the USA, where mass car ownership first caught on, some cities were losing their population to their outer suburbs and the countryside beyond as early as the 1930s. In Britain, counterurbanisation had its roots in the Second World War. So much housing stock was destroyed within both inner and outer city suburbs that the total populations of the metropolitan areas went into decline, and generally have never recovered. The building of the New Towns to absorb overspill population also played a great role in the counterurbanisation process. London reached its peak population in 1939 when it was 8.6 million; in 1951 it had declined to 8.2 million. In the next three intercensal decades the population continued to fall: by 2.5% between 1951 and 1961, by 6.8% between 1961 and 1971 and by a remarkable 9.9% between 1971 and 1981. Although London has undergone some **re-urbanisation** in the past few decades through gentrification of inner suburbs and redevelopments such as that in Docklands, it has not yet regained its pre-war population.

Hand in hand with counterurbanisation is the process of **suburbanisation**. This term has two meanings: at one level it is simply the outward growth of suburbs as they eat their way into the countryside on the urban fringes. The other meaning of suburbanisation is the whole range of changes that take place as people from cities move into the outer suburbs, the market towns and the villages of the countryside around cities. These migrants change the countryside economically, socially, demographically and culturally. Figure 7.1 illustrates these changes in the context of Britain, but generally they are applicable to the situation in other MEDCs. Some of the most profound changes include:

- ageing of the population through retirees settling in the countryside
- rising property values that cause problems for established local families on low incomes (especially young couples buying their first homes)
- closure of services such as village schools, shops and bus services, as they are less likely to be patronised by newcomers
- greater politicisation of green issues, as newcomers bring with them much more Nimbyist attitudes.

Despite the fact that London and other cities in Britain are experiencing some re-urbanisation, counterurbanisation is still the dominant migratory demographic change within the country, and by the first few years of the twenty-first century the rate of movement

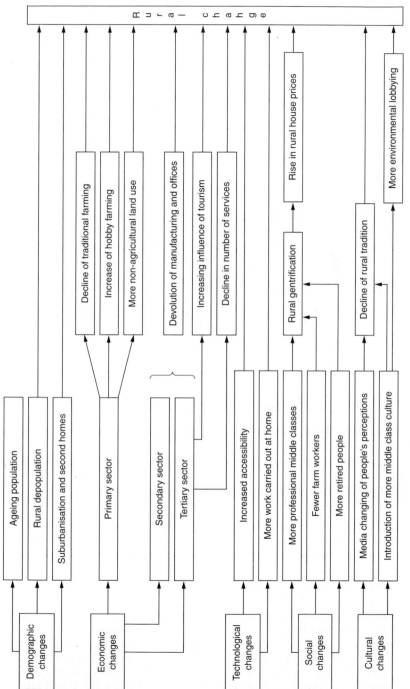

Figure 7.1 The processes associated with suburbanisation

from cities to rural areas averaged 90 000 people per year. Also, a survey carried out at the beginning of the new millennium showed that 51% of the urban population of the UK would prefer to live in the countryside, if they had the opportunity.

2 The Development of the Edge City

The **Edge City** is a type of outer suburb that began to evolve in the USA in the 1980s as a result of increasing counterurbanisation and suburbanisation. Although the concept of having decentralised ideal communities and places of work on the city edge and beyond (effectively in the British Garden City tradition) had been conceived in the USA by the architect Frank Lloyd Wright in the 1930s, when they were known as **broadacre cities**, very few were actually built.

As counterurbanisation proceeded, people moved further and further away from city centres and settled on the outer urban fringes. Then the shopping malls and other services soon followed and located in these edge cities. Next came the places of employment, the glittering office blocks housing tertiary industries and the landscaped factories of high-tech industries. All of these functions: residential, retailing, financial and other services, as well as high-tech factories, are low-rise and land-extensive, taking up vast tracts of green spaces on the urban fringe. Joel Garreau produced the definitive work on these far outer suburbs of US cities, entitled *Edge Cities: Life on the New Frontier*, in 1991. In this he both defines what is meant by an Edge City and gives a detailed description of its character (or lack of it). Some of the main criteria Garreau set for Edge Cities were:

* They should have at least 5 million square feet of leasable office space.
* They should have at least 600 000 square feet of leasable retail space.
* They should have more jobs than bedrooms.
* Their population should perceive them as a separate and individual place.

Garreau identified two main types of Edge Cities: the '**Uptowns**' which were formed around a pre-existing smaller towns on the urban fringes (in sharp contrast to the Downtowns of the city centres), and the more rapidly growing '**Boomers**' which were located at nodal points on expressway (motorway) networks. The way in which new, mall-based settlements locate themselves at strategic crossroads on motorway networks has given rise to the term **mallopolis**. Typically, a mallopolis is located between 30 and 100 km from a city centre. The spread of these types of outer suburbia has been such that now over 60% of office space in the USA is in the suburban rather than CBD locations. For example in Houston, Texas, one Edge City, Post Oak, alone has 62.5 million square feet of office space compared with just 38 million square feet located downtown. The way in which offices relocate from CBDs straight to the city edges is often described as **leap-frogging**: a reference to the fact that the relocation process misses out the various belts of inner and middle suburbs which lie in between.

Soja (2000) regarded the evolution of Edge Cities as 'the urbanisation of suburbia' and that it is an integral part of the transition from the industrial to the post-industrial city. Where an Edge City particularly specialises in high-tech industries it may be located and even integrated with a university campus, as is the case with North Carolina State University at Raleigh. Manuel Castells and Peter Hall (1994) coined the term **technopole** for this type of settlement. In the USA, Kotkin, in his work *The New Geography* uses the term '**nerdistans**' for these settlements, which he describes as:

'... self-contained high-end suburbs that have grown up to service the needs of both the burgeoning high-technology industries and their workers ... Successful nerdistans seek to eliminate all kinds of distractions – crime, traffic, commercial blight – that have commonly been endemic in cities ...'

He also suggests that nerdistans are not true Edge Cities as they in no way depend on the city core and do not duplicate the traditional functions of the CBD.

The inhabitants of the typical Edge City still commute by car, but travel relatively short distances on fast new road networks, and avoid other parts of the metropolitan areas of which their community is a satellite. These new roads have destroyed yet more of the countryside and public transport tends to be minimal because the low density of population is unable to support it, as well as the sheer areal spread of the community. At the heart of the Edge City, as Pretty (1998) writes: 'the new monument – no longer a cathedral – is the atrium, a climate controlled environment for the shopping experience'.

The Edge City is a place in which it is very practical and convenient to live, yet it has neither the benefits of city centre living nor those of life in the real countryside. Also, as they are new settlements with the anonymity of the suburbs, Pretty labels them as 'Places with no History'. Over 200 Edge Cities have been identified as such within the USA. Around greater New York there are already 21 Edge Cities, but it is in southern California, where the concept has become most developed. Within a radius of 100 km of Los Angeles there are 26 Edge Cities, and their highest concentration is to be found in the wealthy Orange County. Some commentators have likened the more extravagant of these communities to theme parks.

Generally Edge Cities are safer for their inhabitants than the suburbs closer to city centres, and gated communities are common. By moving to Edge Cities, however, people do not necessarily avoid the problems, real or perceived, that they leave behind in the metropolitan area. Some of the more remote communities, which have been labelled 'Off-the-Edge Cities' by Soja, do not have the range of employment and services that other Edge Cities attract (and, to use one of Garreau's criteria, certainly do not have more jobs than beds). One such place is the fastest growing of Los Angeles' outer satellites, Moreno Valley, which lies some 100 km to the east of Downtown LA. This has few places of employment nearby and many of its inhabitants spend more than five hours commuting; this has had a knock-on effect within the society of Moreno Valley. The community suffers from above average rates of suicide, juvenile crime, burglary, drug addiction and domestic violence.

Edge Cities are beginning to emerge in other parts of the world, but not always in the same way as in the USA. Within Europe, a group of outer commercial districts on the edges of certain capital cities formed in 1995 the European Edge Cities Network, which shares common concerns as well as the practicalities of development and change. The towns and boroughs that belong to this network include Croydon (London), Nacka (Stockholm), Espoo (Helsinki), Ballerap (Copenhagen), Getafe (Madrid) and Kifissia (Athens). Kifissia, a wealthy suburb on the northern fringe of Athens, particularly gained importance with the Olympic Games of 2004 and the related extension of the Athens metro network. Croydon, with its 333 000 population, provides the biggest contrast with the Californian model. Croydon was a historic market town that was absorbed into the urban fabric of the London metropolitan area by the 1930s. It is therefore new neither as a settlement nor as an outer suburb. However, it does have one similarity with US Edge Cities in the nature of its range of functions. Of the 9000 companies in Croydon, 90% are **SMEs (small and medium sized enterprises)**, of which 82% are service industries (including 29% involved in the financial services sector) and only 12% are involved in manufacturing.

In Britain, there is as yet no real equivalent to the outer fringes of the southern California metropolitan areas. Several commentators, however, such as Marion Shoard in her chapter in Jenkins' book, *Remaking the Landscape*, entitled 'Edgelands' draws our attention to the northern fringe of Bristol. Only known as the 'Northern Fringe' this is an Edge City in the making. With the capacity to employ over 60 000 people, and rivalling the CBD of Bristol, it has a vast out-of-town shopping centre (Cribb's Causeway), together with new Ministry of Defence offices, employing over 6000 people, a science park (Aztec West) and the British or European headquarters of several major high-tech companies. Among all these amenities are huge areas of low-density housing, all with easy access to both the M4 and M5 motorways.

3 The Edgeless City

Planners and urban geographers have widely expected Edge Cities in the USA to mature into more distinctive urban settlements, with higher building densities, more functions and connections to mass transit rail or tram networks, and to become generally more pedestrian-friendly environments. It is becoming clear that this is not happening in many cases, and the land on the edges of large metropolitan areas has just continued to be built on by amorphous low-density sprawl. The urban demographer Robert F. Lang coined the expression **Edgeless Cities** for such settlements. In his 2003 publication *Edgeless Cities: Exploring the Elusive Metropolis*, Lang argues that whereas the development of true Edge Cities could well be slowing down, Edgeless Cities are expanding throughout the USA. He suggests that this represents the main type of twenty-first century decentralisation that is taking place, and that it is a 'post-polycentric' phase in urban development in the sense that Edgeless Cities are not really nucleations but merely a form of sprawl strung out along major highways. Even though they have shopping malls and office blocks, they cannot be regarded as 'centres' in any sense of the word. As Lang writes:

> 'Edgeless Cities are not Edge Cities waiting to happen. Instead, they represent a concurrent, competing and more decentralised form of office development.'

Large US metropolitan areas which have considerable 'edgeless' developments around them include Atlanta, Boston, Chicago, Dallas, Denver, Miami, New York, Washington and, of course, Los Angeles. Lang argues that these developments are never likely to pose a threat to downtown areas as they lack the ethnic diversity and the range of subcultures found closer to city centres, which are so important in such growth activities as entertainment, fashion, advertising, publishing and the media in general.

4 Land-use Changes on the Urban Edge

The land use of the urban fringe may well be associated with the functions of Edge Cities in the USA, but elsewhere the picture is often quite different. In Britain and other European countries, the outer city edge is often blighted and has a range of land uses that are there in anticipation of later developments. Among some newer buildings the land-owners either totally neglect the sites or rent them out to short-term functions. One of the acronyms given to such neglected areas is '**Toads**' (Temporarily Obsolete Abandoned Derelict Sites).

Marion Shoard, who labelled the unorderly urban fringes of British cities '**edgelands**', describes them thus:

'Between urban and rural stands a landscape quite different from either. Often vast in area, though hardly noticed, it is characterised by rubbish tips and warehouses, superstores and derelict industrial plant, office parks and gypsy encampments, golf courses, allotments and fragmented, frequently scruffy farmland'.

The situation is much the same in the southern parts of continental Europe where city edges have continued to grow rapidly in recent decades as a result of both counterubanisation and the transition from extended to nuclear families. The Italian architect Aldo Rossi has given such areas the epithet *zone amorfe* (shapeless zones). An example of this can be seen by anyone travelling by road from Palese Airport to the centre of Bari, in Puglia, southern Italy. They would notice a wide range of functions on the city edge – which are repeated throughout much of Mediterranean Europe. Elements within this type of landscape include:

- farmhouses still occupied, with well-tended fields of Mediterranean tree crops attached to them
- derelict farmhouses (some occupied by squatters) with overgrown fields attached to them
- broken-down dry stone walls littering former neat, carefully tended farmland
- fields used as rubbish dumps, especially for discarded building materials
- temporary industrial units storing items such as building materials and motorcar spare parts
- garages and motor mechanics
- construction sites for new blocks of flats, industrial units and new infrastructure
- abandoned construction sites where buildings are illegal or constructors have run out of funds
- hypermarkets and other forms of low-density retail outlets
- wasteland let to advertising companies to display their billboards.

The city edge landscape may be even more extreme in the context of LEDCs, where urbanisation is still taking place and villages on the urban fringes are either being swallowed up by the expanding city itself or undergoing change as a result of the demands of the urban economy. These demands include both food supplies and building materials and certain studies in India have considered the rural–urban relationships of the edges of large cities.

5 Green Belts and the Protection of the Urban Fringe

The most successful form of remedial action against urban sprawl and the blight caused by edge-of-city land uses is the **green belt**. The concept has its origins back in the utopian planning tradition of such people as Ebenezer Howard and the Garden City movement. In Britain, the first green belts were set up by the Town and Country Planning Act of 1947. Today there are 15 well-established green belts, by far the largest of which is the 1.5 million hectares which surrounds London.

Green belts have been a qualified success in Britain and the London green belt has been a good example of the successes and failures of the concept. On the positive side, London has benefited in the following ways:

- The green belt has protected a great deal of farmland.
- The distinction between London itself and the towns beyond the green belt has been retained.
- The belt has had some success in limiting commuting because of that distinction.
- More recreational land is available to the London population than if the belt had not been created.

On the negative side, too many breaches have been made in London's green belt, and these have limited its effectiveness:

- Most of the M25 has been built within the belt and places of great environmental importance have become degraded, e.g. the North Downs Way between Caterham, Surrey, and Westerham, Kent.
- There is so much pressure on greenfield sites for housing; four million new homes are needed in Britain by 2016 and 25% of these are to be built in the south-east.
- Socioeconomic issues may outweigh environmental ones, e.g. the building of new hospitals or new high-tech research institutions may take priority over conservation.
- The management of some parts of the green belt may be poor, leaving it unattractive to locals and city dwellers alike; some parts may even become degraded into 'edgelands'.

Questions

1. Explain the main differences between **counterurbanisation** and **suburbanisation** and the impact they have upon the countryside around towns and cities.
2. Explain how the different types of 'out of town' land uses on the edges of cities evolve and what problems they pose to their inhabitants.
3. Why are the types of land use on the edges of cities so often so diverse and yet poor in landscape quality?
4. What are the advantages and disadvantages to people and the environment of the development of ever-increasing commuting distances?

8 Urban Change in the Changing World Order

> 'The modernity of the liberal West is difficult to achieve for many societies around the world.'
>
> Francis Fukuyama *State Building*

Fukuyama, in his most famous work, *The End of History and the Last Man*, regarded the end of the Cold War and therefore the end of the **Old World Order** as the finishing point in world history. He suggested that the defeat and collapse of Communism by the early 1990s meant that the whole world would follow the Western model of capitalism and liberal democracy. It has become very clear during this last decade and a half that this is not happening and that the **New World Order** is probably creating a more fragmented world. The cause of friction within the world is no longer the capitalist–Communist divide, but rather a whole series of different ideologies and levels of wealth. Fukuyama now acknowledges this in his most recent publication and that is reflected in the quotation above.

The Old World Order had the rather oversimplified divisions of First, Second and Third Worlds. The First World referred to the capitalist industrialised liberal democracies of the West, which included the USA, Western Europe, Japan and Australia. The Second World was made up of the Eastern European bloc of industrialised centrally planned economies led by the old USSR. The Third World (now a term which is both pejorative and geographically incorrect) comprised all the remaining poorer, non-industrialised nations.

1 The New World Order and its Impact on Urbanism

Since the collapse of Communism, a new and much more complex world order has come into being. Rather than just three levels of development, the world has many different subdivisions of nations, best classified by using the categories of the World Bank, with a few slight modifications:

- The **OECD** (Organisation for Economic Co-operation and Development) countries: these are the old First World countries (the USA, Japan, the countries of Western Europe, Australia, Canada, New Zealand, but the group is gradually expanding as new nations such as Korea, Mexico and Poland have joined in recent years).
- The **Transition Economies**: these are the former Communist countries of Eastern Europe which are undergoing the transition from centrally planned to capitalist economies. Many of these joined the European Union (EU) in 2004, but have not yet qualified to join the OECD (e.g. Latvia and Lithuania); others such as Belarus and Russia as yet have no aspirations to join the EU.
- The **Oil-exporting Countries** (e.g. Saudi Arabia and the United Arab Emirates): many of these have greater wealth than some OECD countries.
- The **Newly Industrialised Countries** (NICs): these may also have wealth comparable to that of OECD countries; they include Singapore, Taiwan and Hong Kong and are known collectively as the Asian 'Tiger Economies'.
- The **Newly Industrialising Countries**: these include Asian countries following the 'Tiger Economies' model (e.g. Malaysia and Thailand) as well as booming economies in other parts of the world, such as Brazil.
- The **Higher Middle Income Countries**: the more successful economies in LEDCs which have yet not reached the NIC stage fall into this category, e.g. Tunisia and Vietnam.
- The **Lower Middle Income Countries**: within this category lie the bulk of LEDCs, e.g. Peru, Egypt, Ghana and Burma.
- The **Low-income Countries**: these are the countries whose development has been held back by ruthless regimes, civil wars, natural disasters, famine and disease, e.g. Afghanistan, Somalia, Zimbabwe, Sudan and Sierra Leone (the majority of countries in this category lie in sub-Saharan Africa).

Where countries lie within this pattern of the New World Order affects not only their income, but also how their cities are developing and changing. OECD countries, along with the oil-rich nations, have the greatest opportunities to develop their cities in a well-planned way. The two types of NICs generally have the fastest growing commercial sectors and therefore are likely to have the most rapidly changing CBDs. By contrast, the transition economies are likely to be making the greatest overall adjustments to their city layouts, by investing in the commercial upgrading of their centres, building new private-sector housing and having to upgrade their industrial zones.

Cities within countries of the middle income brackets are those which are still experiencing very high rural–urban migration and therefore housing is often the single most important issue facing city planners. At the same time, the huge daily movements of people to and from work make transport and congestion very important issues too. The low-income countries also generally suffer from the problems of housing shortages and traffic, but if they are recovering from war or natural disasters, they may also be facing other issues such as influxes of large numbers of refugees and the disruption of basic services such as electricity and water supplies.

The 2004/2005 UN-HABITAT report *The State of the World's Cities* makes the differences between the regions that include LEDCs very clear. It states that for Asia, in general, levels of poverty have declined from over 50% of the population in the 1970s to less than 25% today. This has had a great impact on the quality of life of urban dwellers. In Latin America, however, the picture is far from rosy. Although the 1970s and 1980s were generally boom years, following the recession that started in 1997, many economies have gone into great decline. This has seen not only an increase in urban poverty, but also greater social polarisation of rich and poor neighbourhoods. It is above all sub-Saharan Africa that is suffering from severe urban problems, with 166.2 million people or 72% of the total of urban residents living in slum conditions.

This chapter will now look in detail at some of the most important categories in the New World Order and cities that have characteristics typical of them.

2 Urban Development in Oil-rich Economies

Oil-rich economies have been able to invest vast amounts of money in infrastructure and therefore urban development in the past few decades. In the Middle East, it has been the smaller oil-rich states, with their relatively small populations and territories, that have benefited and changed most. The four Gulf states of Kuwait, Bahrain, Qatar and the United Arab Emirates (UAE) are all examples of economies which have grown rapidly, and had oil revenues wisely invested, creating economic diversity in preparation for the time when either the oil runs out or it is superseded by other fuels.

CASE STUDY: DUBAI

Dubai, the capital of the Emirate of Dubai (one of the seven states that make up the UAE) is probably the fastest growing city in the Middle East. The UAE was the most successful of the small Gulf states in diversifying its economy. Oil now accounts for less than 40% of the country's income. Most of its new activities are in the tertiary sector, and nowhere is the economic boom felt more than in Dubai.

Dubai was until the twentieth century a relatively small port on the Gulf that specialised in pearl fishing, gold smuggling and trade across the Arabian Sea. It was the enlightened ruler Sheikh Rashid who, from the 1930s to the 1980s, set out to expand Dubai's trade to make it one of the main ports of the Gulf. The real boom came from the 1970s onwards when oil income could be invested in large-scale projects. Liberalisation of the economy led to the setting up of the major free port zone and the inward investment by a wide range of international corporations in the industrial, commercial and other service sectors.

Today, with a population of 1.5 million, Dubai is often referred to as one of the world's biggest construction sites because so much development is taking place there. The city and Emirate are taking advantage of its geographical position within the Middle East, with its long sandy shoreline, clear seas and coral reefs, its warm winter climate and its focal point on air routes from Europe to Southeast Asia and Australasia, to make it into one of the world's most important tourist destinations. As a regional and international leisure, recreational and entertainment centre, some of its main tourist facilities include:

- many world-class hotels, including the Burj El Arab high-rise tower
- resorts and residential areas on newly reclaimed land
- resorts based on individual sports such as golf, tennis and scuba diving
- theme parks, many with a strong Arabic and Middle Eastern flavour
- resorts and activities using the desert beyond the city's edge.

The current and future urban structure and land use in Dubai are illustrated in Figure 8.1. The developments are taking place in a cellular manner and there is considerable segregation of land use.

The original core of the settlement is represented by the historical fort and port area of Bur Dubai on the southern shore of Dubai Creek. On the northern side of the Creek is the main CBD area, Deira, with its bazaars, banks, commercial offices, shops, restaurants and hotels located close to the international airport. Dubai can no longer develop in this direction as its boundary with Sharjah Emirate is close by.

The rapid development of the city has therefore been taking place mainly southwards along the coast, as well as inland into the desert. In the 1980s and 1990s, the main tourist and residential buildings were constructed along the Jumeirah Beach area

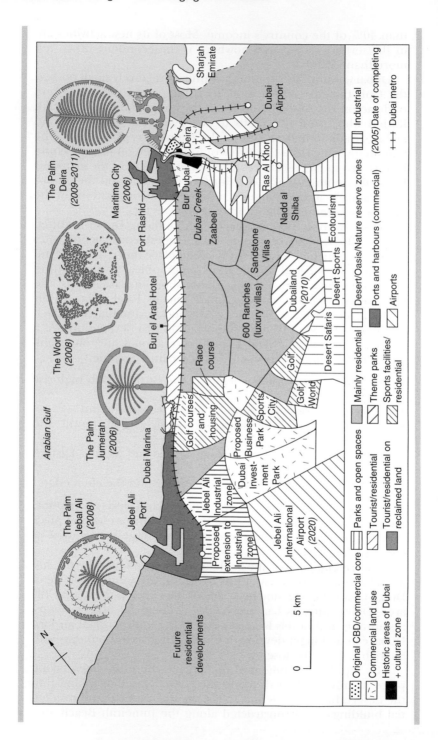

Legend:
- Original CBD/commercial core
- Commercial land use
- Historic areas of Dubai + cultural zone
- Parks and open spaces
- Tourist/residential
- Tourist/residential on reclaimed land
- Mainly residential
- Theme parks
- Sports facilities/residential
- Desert/Oasis/Nature reserve zones
- Ports and harbours (commercial)
- Airports
- Industrial
- (2005) Date of completing
- +++ Dubai metro

between Bur Dubai and the new Jebel Ali Port. This huge deep-water container port has become the hub of Dubai's heavy and light industrial development. Dubai's trading, industrial and commercial activities are transforming it into a Middle Eastern equivalent of Hong Kong or Singapore.

Jebel Ali is now the centre of Dubai's built-up area and is a zone of luxury housing, with important sporting facilities such as the Emirates Golf Club, the main university campus, Internet City and a whole string of luxury hotels along the beach. The Dubai government has a very liberal policy on foreigners owning property, which has stimulated both economic development and urban growth (it is said that all of the English football team own properties in Dubai). In the Jebel Ali district, the first massive land reclamation scheme has taken place. The Palm Jumeirah is one of four planned tourist and residential areas – three of which will be in the shape of palm trees, the fourth a map of the world. The new housing is in a wide variety of styles. The Nakheel project known as the 'Lost City' is a romanticised and idealised vision of the past. Built on 560 hectares of land amidst folly-like ruins which evoke past civilisations of the Middle East and North Africa, it is divided into five 'village' communities, each in a different traditional architectural style: Iraqi, Syrian, Moroccan, Palestinian and Lebanese. In this luxury development the boundaries between fantasy and reality will indeed be blurred.

The newly created beaches, luxury hotels, and private flats and villas will give Dubai – which had 5 million visitors in the first half of 2005 – the greatest tourist capacity of anywhere in the world. The huge north–south axis of developments will be linked by a light rail metro system by 2007. With the southward shift of developments and the limited area of the existing airport, the new Jebel Ali International Airport is planned to be built between 2006 and 2020.

However, not all of the changes are turning Dubai into asphalt and concrete; some areas of desert on the eastern fringe of the city are being incorporated into the urban plan and becoming centres of conservation and ecotourism.

With its wide range of eclectic and high-tech styles of architecture, its rapid growth as a centre for leisure industries, and the vast sums of money being invested in these changes, is becoming Dubai the most remarkable post-modern fantasy city on earth; it is already eclipsing places in the USA such as Palm Springs and Las Vegas in this role.

opposite **Figure 8.1** Dubai: land use and development

3 Dynamic Urban Change in Newly Industrialising Countries

Although not as dramatic as the changes that are going on in the oil-rich countries, the cities of the newly industrialising countries are undergoing rapid growth and a transformation that is making them increasingly like urban areas in MEDCs. Cities within newly industrialising countries are, to some extent, able to learn from the mistakes of MEDC cities and adjust more quickly to the needs of the modern world. On the one hand they may suffer from the problems of traffic congestion and squatter settlements, but on the other they lack the problems of derelict industrial plant dating back to the nineteenth century. The most successful cities have tended to be those where liberal policies from national governments have allowed foreign firms easy access, which has led to high levels of investment and technical know-how in the development of infrastructure, industry, housing and commercial services.

CASE STUDY: BANGKOK, THAILAND

Thailand, with a population of 60 million, is one of the most rapidly urbanising countries in the world. The urban population accounts for 22% of the total, and just over half of these people live in the capital, Bangkok. With its fuller name, Krung Thep Maha Nakon, Bangkok was for centuries the second city of Siam, until the capital was transferred there in the eighteenth century. The most rapid growth of this Asian megacity took place in the second half of the twentieth century when it went from 700 000 people to the present 6 million within the city administrative area and 8 million within the Greater Bangkok area. Originally built along a series of canals, the biggest problem the city faces today is its traffic congestion as a result of its transition from canal to road transport.

Figure 8.2 shows certain aspects of the structure and development of Bangkok. The historic core of the city is the area close to the Chao Phraya River, the Royal Place complex and a series of important temples. This was the original administrative and commercial district. Today it still has some ministries, as well as some older manufacturing districts such as the timber warehouses and furniture workshops along the *klong* (canal) at Damrong Rak, the wholesale markets such as the clothing and textile market at Bo Bae, and the huge traditional retailing Chinatown district, with different streets specialising in particular wares.

The current CBD of Bangkok is located in two ever-changing high-rise areas separated by a few square kilometres of low-rise

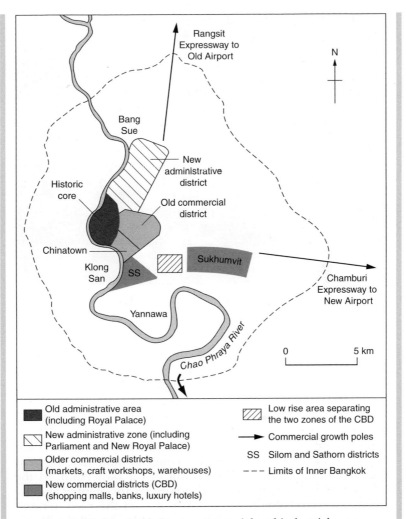

Figure 8.2 Bangkok's inner commercial and industrial zones

buildings and parks. The Silom–Sathorn zone has the main banks and financial institutions, many of the main hotels, some shopping centres and the entertainment (including the red-light) district. The Sukhumvit–Siam Square zone has the main multi-storey shopping centres, many company headquarters and numerous luxury hotels, as well as being the main diplomatic district. Both of the CBD zones have experienced a great deal of intensification in their land use, building height and building densities as pressure on space and land values have led to both

Central Bangkok, Thailand

vertical development and infill. Traffic movement between these two zones was very slow until the opening of the elevated 'Skytrain' metro link in 1995. The administrative part of the city now lies to the north of the historic core, and unlike the two CBD zones, is fairly low density with considerable areas of parkland including Bangkok Zoo.

The rapid expansion of Bangkok in the last few decades was facilitated by the construction of a complex multi-lane highway network accompanied by the growth of car ownership. In the centre of Bangkok many of these motorways are elevated and the cause of considerable pollution. Two axes of the road network are determining present and future growth; the northern axis out to the current Don Muang Airport and the eastern axis which leads to the deep water port at Chon Buri as well as the new Suvarnabhumi Airport, due to open in 2007. Along both of these axes are the main national and foreign-owned industrial plants, high-rise mini-CBDs, hotel, conference and trade fair complexes. They are located here because of their relatively high levels of accessibility.

Motorways have also been responsible for the evolution of vast areas of suburban low-rise housing estates known as *muban jatsan*. The wealthy and the new middle classes either live in these peripheral areas and commute into Bangkok or live in more central condominium blocks of flats. The poorer districts are scattered throughout the city and include some of the traditional *klong*-side villages, the medium-rise inner city and mid-suburban districts, city centre areas between high-rise developments and around peripheral villages where new developments have not taken place. With its great desire to have the image of a modern, developed city, the Bangkok authority has invested greatly in upgrading housing, self-help schemes and the laying on of public amenities in the poorer districts, which is transforming them to an acceptable standard. By as early as 1980, the proportion of poor housing had dropped below the 25% level, and a process known as **land sharing** was in operation; this led to the upgrading of poorer districts, although it also increased building and population densities. Some of the *klong*-side settlements in central Bangkok, which a few decades ago were heavily polluted and lacked basic amenities, now have an almost gentrified appearance as a result of individual action through self-help as well as local government intervention.

Since the mid-1990s, Bangkok has been overcoming its poor public transport network with the opening of its first metro lines. In 1995 the first two elevated urban railway lines of the Skytrain were opened, linking the Silom and Sukhumvit Roads. In 2004 the first underground line was opened and this linked the central station with various parts of the city centre and with various Skytrain interchanges. In February 2005, when Prime Minister Thaksin was re-elected for a second term of office, he announced that he would go ahead with the expansion of the rapid transit network to cover a comprehensive 248 km, reaching most parts of the city and linking it to both the old and new international airports. The seven lines will be run by three separate companies; on the existing network there are two companies and as yet they lack an integrated ticketing system. After decades of haphazard development and traffic chaos, the basic urban structure of Bangkok appears to be coming under much closer control.

4 Urban Change in Transition Economies

Although many of the cities in LEDCs and NICs are undergoing unprecedented growth and are consequently having to cope with the many aspects of rapid urbanisation, the cities of the former Communist states are also experiencing much more radical economic and structural changes. Before the collapse of the Communist bloc in Europe and central Asia, the cities of these centrally planned economies were distinctly different from Western cities, much more than they are today, and they often warranted separate chapters in

urban geography texts. Although many of these cities have undergone very rapid change in the decade and a half since the fall of the
Berlin Wall, many are still left with vast Soviet-style residential areas
and industrial zones.

The mould for the typical Communist city was effectively created
when Moscow's 1935 master plan was devised by Soviet planners. The
basic principles were adopted and copied all over Eastern Europe.
Some of the most important of these principles were:

- to fix a limit to the population of the city and then control internal
  migration
- to have total state control over housing, its unit size, density and
  facilities provided
- to subdivide the city into neighbourhoods of similar size
  (8000–12 000 people in the case of Moscow)
- to create monumental zones in the city centre to instill in people a
  sense of patriotism (also for purposes of Communist propaganda);
  part of this zone would be taken up by huge squares and avenues
  for state-initiated parades and marches
- to have extensive areas of open space for recreation and leisure
- to limit journeys to work by having sufficient employment in each
  neighbourhood
- to have strict land-use zoning within neighbourhoods and within
  the city itself
- to rationalise traffic flows by encouraging heavy traffic to use wide
  boulevards.

What has been happening in the transition economies in the last
two decades has been the dismantling of some of the old urban fabric,
replacing it with new buildings and new functions. Many of the historic cores of the cities of former Communist countries were of great
beauty, but with a rather run-down feel. In such places restoration of
buildings, the creation of new privately owned hotels, bars and restaurants and the development of the tourist industry have often brought
about rapid change. In the early 1990s, Prague became the first
Eastern European capital to embrace mass tourism from the West,
but in the following decade such cities as Tallinn (Estonia), Bratislava
(Slovakia) and Ljubljana (Slovenia) have undergone this process of
'**Prague-isation**'.

Some of the greatest changes that the cities of transition
economies are witnessing are:

- the demolition of some of the monumental buildings, or their
  adaptation to other functions (e.g. the conversion of Hoxa's
  planned mausoleum in Tirana, Albania, into a trade fair exhibition
  hall, mentioned later in this chapter)
- the setting up and operation of transnational corporations (TNCs)
  and foreign banks and the construction of new prestige office
  blocks in the CBDs to house them (e.g. the World Trade Centre

chain of office blocks throughout Eastern Europe, which has helped to revive the commercial centres of in such cities as Iaşi, Romania)
- the conversion of the main state-owned stores to Western-style department stores (e.g. the ZUM store in Bishkek, Kirghizstan)
- the setting up of shopping malls (such as the extensive one underneath Victory Square in Kiev)
- the restoration or demolition of decaying and substandard housing units
- the development of new inner city and suburban middle class and luxury housing for the new rich; such areas would have been the preserves of the Communist Party officials in the past (e.g. the southern suburbs of Almaty, in oil-rich Kazakhstan, which stretch uptowards the foothills of Altay mountain range).

At the same time as these dynamic changes are taking place to the urban fabric of former-Communist cities, there have been considerable social changes, particularly in the polarisation between the rich and the poor. The introduction of a free-market economy has fuelled economic growth and numerous families have made their fortunes from being the entrepreneurs of the transition process. On the other hand, the closure of obsolete heavy industries has led to much higher unemployment rates and poverty levels. Levels of poverty have been further exacerbated by the very high inflation rates that these economies have experienced during the transition period. Only with the passing of time will it be clear whether the transition from Communism to a free-market economy has been to the benefit of the many rather than the few.

Sasha Tsenkova of the University of Calgary, who uses the term **post-socialist cities** for cities in transition economies, carried out analytical research on Prague, Riga, Budapest, Ljubliana, Vilnius, Belgade, Zagreb, Tallinn, Sofia and Tirana, and found that whereas the economic transition in many places was fairly swift through changes such as privatisation, the social transition has been less rapid. Unemployment, retraining for new occupations, affordable housing and access to services have become major problems facing large numbers of urban dwellers and have consequently led to increasing social polarisation. However, people are likely to be better off in capital cities than in secondary cities as capitals generally attract much more foreign investment and also more income from tourism.

CASE STUDY: TIRANA, ALBANIA

Albania was the hardest of all the hard-line Communist countries. From the end of the Second World War until the collapse of Communism in the early 1990s, it systematically isolated itself from all its potential allies: Yugoslavia, China and the USSR, and remained cut off from the outside world. Under its ruthless dictator Enver Hoxha it became the most secretive and economically backward nation in the Eastern bloc, with the government deliberately keeping its people in a state of paranoia and fear of invasion from outside. Both the countryside and the edges of the cities are consequently littered with rows of unattractive concrete bunkers.

Tirana, Albania's capital, which has a population of 800 000 (about 37% of the national total) is located in a basin close to the Dajti Mountain to the east and some 40 km from the Adriatic Sea. The city is emerging well from its period of isolation and Communism, and what is known as the 'Tirana model' is being studied and emulated elsewhere. Keen to get Albania out of the economic depression and political turmoil that followed the collapse of Communism, many Western governments have invested money and expertise to help solve the urban problems of Tirana.

During the decades following the Second World War many of the historic buildings (including the Albanian Orthodox Cathedral and the main bazaar) in the city centre were demolished to make way for drab, functional modern concrete blocks such as those housing the National Historical Museum and the Opera House surrounding the main square (now known as Skandebeg Square after Albania's fifteenth-century national hero). To the south of this square are the various ministries and government buildings. During the Communist period (1945–1991) this zone became a 'forbidden city' for Albanian citizens and only party officials were allowed entry. With the collapse of Communism and the liberalisation of the economy, this part of the city centre has undergone very dynamic changes. New shopping centres, restaurants and bars are being developed in the zone, a Roman Catholic Cathedral was opened in 2000 (under Communism Albania was officially an atheist state) and the pyramid-like building intended to be the mausoleum of Enver Hoxha was converted into an exhibition hall for trade fairs. Drab concrete high-rise buildings of the 1960s and 1970s in this area have been given makeovers using vibrant colours on their façades.

The strategic plan, which has been adopted by Tirana's authorities (see Figure 8.3), is based on seven previous studies carried out using funding from sources as diverse as the World Bank, USAID, the Austrian government and a Rotterdam planning institute. The plan of 2002 has a series of concentric rings around the city centre defined by the road system. From here, a combination of linear and cellular development is taking place north-westwards along the Tirana valley towards Mother Teresa International Airport (recently reconstructed with Canadian

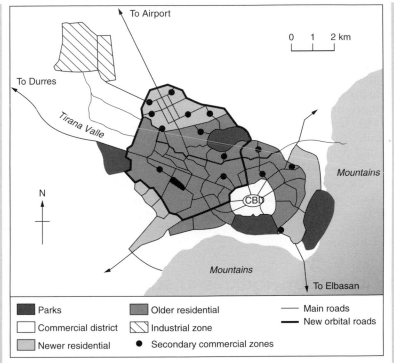

To Airport

0   1   2 km

To Durres

Tirana Valle

Mountains

N

CBD

Mountains

To Elbasan

■ Parks   ■ Older residential   ——— Main roads
☐ Commercial district   ◫ Industrial zone   ▬▬▬ New orbital roads
■ Newer residential   ● Secondary commercial zones

Figure 8.3 The Tirana development plan

support). The city cannot develop very far eastwards or south-wards because of mountains and hills.

It is in the Tirana valley that most development and reconstruction is needed. This zone is highly degraded with poor-quality housing, ranging from rural cottages which have been engulfed by the growth of Tirana, to cheaply built and badly maintained high-rise blocks from the Communist era, and squatter settlements along the river and railway lines. The environment here is both ecologically and visually polluted with vast smouldering rubbish tips, gypsy encampments and endless rows of concrete bunkers dating from the paranoid Hoxha era.

5 Urban Decay in Stagnant Economies

As mentioned above, the cities that are in the worst state of development are those which have been torn by wars (civil or international), have suffered from the long-term effects of natural disasters, or have suffered from corrupt regimes or international sanctions that have dragged them down economically. The longer the period of disruption, the more the people and the infrastructure of the cities they inhabit suffer. It is remarkable how resilient populations can be and how quickly cities can recover, however. In Bosnia-Herzegovina, the cities of Sarajevo and Mostar are undergoing fairly rapid reconstruction following the civil war of the mid-1990s. However, physical rebuilding does not necessarily cure psychological damage and there remains great mistrust between the various ethnic groups who were formerly at war with each other.

The beneficiaries from recovery may not always be the local people. In Kabul, the Afghan capital, the sheer number of foreign United Nations (UN), non-governmental organisation (NGO) and charity workers on salaries much higher than local ones has led to gentrification in some of the residential areas of the city where there are large historic houses. In Autumn 2004 a local London newspaper, the *Islington Gazette*, complained about the way in which its borough was being ridiculed, as Kabul's gentrification had gained it the epithet of 'Islington-on-Kush'!

CASE STUDY: MOGADISHU, SOMALIA

In 2004, Somalia had had no proper government for over 15 years. The country, and its capital Mogadishu, was still in the hands of a series of warlords who had carved out their fiefdoms during the protracted civil war that started in 1991 with the overthrow of the dictator Siad Barre. From 1992 to 1993 civil war and the related problems of famine and food distribution claimed over 300 000 lives. The country has since then effectively been in the hands of various warlords, the population being made up of four major clans and dozens of smaller ones. Despite the intervention of the USA and then the UN, the process of state-building has been held back by continued fighting between the various factions within the country, and within Mogadishu itself. During the 15 years of anarchy an estimated 1–3 million people have left Somalia, several hundreds of thousands of these from Mogadishu, although the city has also attracted refugees from the countryside. No census figures are available because of the lack of government.

The city shows the typical signs of a place ravaged by protracted war. The infrastructure is particularly poor. The international airport remains closed, streets are potholed and flooded during the rainy season. Electricity supplies have not been restored and therefore the lack of traffic lights contributes further to traffic chaos. Telecommunications wires are overloaded and frequently break down in the absence of an operating national telecom company. There is no effective police force and the various clans have rival checkpoints throughout the city.

The biggest division is between the northern and southern parts of Mogadishu; to most people the north is a no-go area. With no civil service in operation and education operating only at a basic level, the ministries, the university and various colleges have become large refugee camps.

The human will to survive generates economic activity and many of the city's shopping areas and markets thrive, against all the odds. The main Bakara market is the centre of economic activity and goods are cheaper than in surrounding countries because the lack government means that nothing is taxed. Those who have been able to take most advantage of the situation are the wealthier businessmen with enough capital to set up new enterprises. For example, in the absence of the national telecom company, there are now three thriving private ones. The most successful entrepreneurs in Mogadishu seem to be those who can afford both private security guards and private aircraft. In 2004 a new government took office, but it was formed and is still based in Nairobi, Kenya. Mogadishu is still regarded as too dangerous a place for the Somali government to be there.

It will probably be a long time before Mogadishu becomes a vibrant capital city once again. The problems of Somalia have been partially due to the internal conflicts, but also partially due to outside intervention.

Questions

1. Why has the **New World Order** brought about such a wide diversity in the way cities are now developing?
2. Outline the ways in which **post-socialist** cities are developing and how they differ from what they were like prior to the 1990s.
3. With reference to specific examples, explain why some cities are undergoing much more dynamic change than others.
4. Explain why the economies of certain countries stagnate and the range of urban problems this creates.

9 Ideal Cities, Global Cities, High-tech Cities and Future Cities

'The first person I met at Eden-Olympia was a psychiatrist, and in many ways it seems only too apt that my guide to this "intelligent" city in the hills above Cannes should have been a specialist in mental disorders. I realise now that a kind of waiting madness, like a state of undeclared war, haunted the office buildings of the business park.'

JG Ballard *Super-Cannes*

Ballard paints a rather ominous picture of his fictitious futuristic intelligent city, which in reality is a thinly disguised version of the high-tech hilltown of Sophia-Antipolis sited above the French Riviera. When considering the way cities are changing now and how they will be in the future, the edges become blurred between the real urban world, virtual reality and science fiction.

1 'Utopia' or 'Dystopia'? Visions of the Future City

The ideal city in the context of town planning theories was dealt with in some detail in Chapter 2. In this final chapter, the theme has to be explored further and put into a more post-modern and futuristic context. Whereas the vision of the ideal or futuristic city before the twentieth century was found in literature and paintings, one of the greatest

The Fifth Element's view of the future city

modern media for depicting cities has been the cinema. Many twentieth-century films have shown future cities to be high-tech on the one hand, particularly with their means of transport, but on the other hand, have painted a rather grim picture of extreme social polarisation and degradation of the urban fabric.

Fritz Lang's 1927 classic *Metropolis* is probably the most influential of all city films. Set in 2027, it depicts a high-rise city with multi-level transport lines. The small élite live at the highest levels and enjoy a wide range of entertainment and sports, whereas the poor live deep underground and lead a life of drone-like drudgery. The plot revolves around the tensions between these two classes. Perhaps the most important city movie of the last few decades has been *Blade Runner*, Ridley Scott's 1982 film, set in a 2019 Los Angeles. There are flying cars and other revolutionary modes of transport, but the city centre has become so degraded that only the misfits and deprived are left there and speak a babble dialect. All of the more able and mobile live 'Off-World' in a space colony that reflects the concept of the Edge City. In keeping with the growth of TNCs and globalisation, the city is dominated by one company, the Tyrrell Corporation. Other influential films to portray the city of the future include Luc Besson's 1997 *The Fifth Element*, set in a traffic-congested 60-storey New York, and Alex Proyas' 1998 *Dark City*, in which humanity appears have no future and the night is endless, but where the widely advertised Shell Beach resort acts as a promised land.

2 World Cities, Global Cities and Globalisation

In an increasingly urbanised world, cities dominate society more than ever. In the nineteenth century, it was the spread of industrial cities based on the British model that first led to rapid urbanisation in what are today the MEDCs. In the twentieth century various academics recognised and wrote about **World Cities**, i.e. a small number of very large cities that had and continue to have a great influence on the world economy. In the twenty-first century it is the effects of globalisation and the evolution of high-tech and **Global Cities** that are defining the new hierarchy of urban settlements.

a) World Cities

The concept of the 'World City' came into being in 1915 by the British urban theorist and town planner Patrick Geddes in his book *Cities in Evolution*. Geddes defined his concept of World Cities as simply those places in which a disproportionate amount of the world's business is conducted. In the 1960s Peter Hall wrote his classic book, *The World Cities*, in which he explored the reasons for their dominance, the problems caused by their rapid growth and the way in which planners have dealt with these problems. The list of cities that Hall singled out was interesting: London, Paris, Randstad Holland (the conurbation of Amsterdam, The Hague and Rotterdam), the Rhine-Rhur (a conurbation containing some 13 large German towns and cities, the largest of which was Cologne), Moscow, New York and Tokyo. These were still some of the most important industrial cities during the 1960s; by the 1980s, however, when de-industrialisation was taking place, the most important World Cities were financial and commercial rather than industrial centres. Certainly in the early twenty-first century London, New York, Paris and Tokyo are still in the first league of World Cities as they are among the most important business centres, but other cities have emerged to take the places of some of those on Hall's original list.

Friedmann and Wolff, writing in the 1980s, identified a group of World Cities that had become the control-and-command centres of global capitalism. As TNCs and global capitalism had come to erode the barriers of national frontiers, this group of cities, including London, New York and Tokyo, came to form the apex of a world hierarchy of cities. The complex high-tech links between these major centres enable them to dominate business on a world-wide scale.

Thrift, writing in the 1990s, suggested that the important World Cities are, more than anything else, **centres of interpretation** where the key figures are those highly specialised professionals who interpret and predict the way in which the global economy is moving; they include investment bankers, economists, traders, dealers and financial journalists. At one and the same time, these people are a highly mobile international élite, yet so specialised as to be working in almost separate communities within cities. While communicating electronically on a world-wide basis, they still need face-to-face contact within the city environment.

PJ Taylor carried out detailed analyses of the functions of major world financial centres and drew up his 'hierarchy of world cities'; his results appeared in an article in 2001 entitled 'Urban Hinterworlds: Geographies of Corporate Service Provision under Conditions of Contemporary Globalisation'. Taylor classified world cities into three groupings according to their importance in commerce and related activities, and in particular the number of financial and industrial companies which are based in each place; the three groups are:

- **alpha world cities**: Los Angeles, Chicago, New York, London, Paris, Frankfurt, Milan, Singapore, Hong Kong, Tokyo
- **beta world cities**: San Francisco, Mexico City, Toronto, São Paolo, Madrid, Brussels, Zurich, Moscow, Seoul, Sydney
- **gamma world cities**: Dallas, Houston, Minneapolis, Atlanta, Miami, Washington, Montreal, Caracas, Santiago, Boston, Buenos Aires, Barcelona, Amsterdam, Dusseldorf, Hamburg, Rome, Munich, Copenhagen, Prague, Stockholm, Budapest, Johannesburg, Istanbul, Bangkok, Kuala Lumpur, Jakarta, Beijing, Shanghai, Teipei, Manila, Melbourne.

What is clear from these lists is that in the last few decades an increasing number of cities from outside the old, established MEDCs have joined the ranks of world financial centres. It is difficult to define which of these cities can be regarded as truly global, apart from those in the alpha list.

b) Globalisation and the contemporary city

The **globalisation** process which is so profoundly altering our contemporary world, and in particular the major World Cities, is a complex web of interrelated changes. It is taking place in four main areas: economics, politics, culture and technology. In each case it is the activities taking place in the most advanced cities in the world that are bringing about these changes; it is also these same cities that are experiencing the biggest knock-on effects of the globalisation process.

i) Economic globalisation

Although the globalisation of the world economy has been taking place ever since the European voyages of discovery and period of colonisation from the fifteenth century onwards, the rate at which the process has accelerated has been at its greatest in the past few decades. Until the late 1970s, there were great variations in economic systems throughout the world, particularly between the Communist and non-Communist blocs. Even within the capitalist world there were huge variations in the relationships between public and private enterprise in each country. In the 1980s and 1990s, the fundamental changes which took place in both the USA and Europe (first in Britain, then elsewhere) were to reduce the role of the state to a minimum and allow both older, private companies and the newly privatised ones to operate both freely and on a world-wide basis. With the collapse of the Communist system in all but a few countries by the early 1990s, Russia and its former satellite countries came to embrace free-market economies through rapid privatisation, and the dominance of capitalism on a world-wide basis was complete.

Economic globalisation has to some degree brought about what the US sociologist George Ritzer called the **McDonaldisation** of the marketplace. Certain brands have become almost universal as a result of the operation of TNCs and their franchising. Ritzer holds out McDonald's as a model, as the success of these fast-food restaurants in the USA has been based on their efficiency, reliability, predictability and low prices. Their global success has been based on these same factors, together with a more cultural aspect of globalisation – the aspiration of an American lifestyle, especially among the young. Although not as universal, other US restaurant 'brands' which have followed the McDonald's model by locating throughout major cities in the world are the Planet Hollywood and Hard Rock Café chains.

TNCs are not by any means all based in World Cities or capital cities. Economic globalisation can give certain headquarter cities a greater global influence than their population size alone would justify. In the USA, Seattle and Atlanta are two good examples of this. Seattle, in Washington state, with a population of some 560 000, is the 23rd city of the USA on the basis of population size. However, as it is the international headquarters of the Boeing aircraft industry, of Microsoft software and of the coffee giant, Starbuck's, it is a global player totally disproportionate to its size. In a similar way, Atlanta, Georgia, has just 416 000 people, which makes it the 39th US city in terms of population hierarchy. Yet Atlanta is the global headquarters of Coca-Cola and CNN, the world-wide television news corporation. Both companies give the city enormous global influence, and are no doubt part of the reason why it bid successfully to host the Olympic Games in 1996.

Taking this thesis one stage further, there is the city of Rochester, New York State. Although not quite in the same world league as

Seattle and Atlanta, it is nevertheless known in the USA as 'the world's image centre'. Rochester has the headquarters of both the Xerox Corporation and Kodak photographic company, yet with just under 220 000 people, it is the 87th USA city in population rank.

CASE STUDY: LONDON AND OTHER EUROPEAN INTERNATIONAL FINANCE CENTRES WITHIN THE WORLD HIERARCHY

One of the most important consequences of economic globalisation has been the evolution of **International Finance Centres (IFCs)**. These are a very select group of cities where much of the global or regional commercial and financial activity is concentrated. On the world stage New York, London and Tokyo are still the three most important financial centres, but many other cities are emerging. In Europe, Frankfurt is a front-ranking rival to London, whereas in East Asia, Singapore and Shanghai are increasingly emerging as rivals to Tokyo.

Faulconbridge (2004) carried out a quantitative analysis of the role of London within Europe's financial centre network and investigated London's changing position in relation to Frankfurt and the other major IFCs (Paris, Amsterdam and Milan). Two of the main reasons why London's hegemony has been challenged in recent years have been the launch of the **European Monetary Union** (and its currency the euro) and the decision by the EU to locate the **European Central Bank (ECB)** in Frankfurt. The survey concluded that while London has continued to dominate the European financial markets, Frankfurt has grown to become a complementary centre and other cities have developed their financial institutions to act as part of a European network within the global network.

---

ii) Political globalisation

In 1995, the Japanese economist Kenichi Ohmae published the important book *The End of the Nation State*. This examined some of the trends that had been going on since the 1980s. Ohmae dealt with the weakening of the nation state, and the decline of its role in economic intervention (the change from the so-called Keynsian state to the competitive state). At the same time both international politico-economic blocs and TNCs have continued to grow in importance.

Political globalisation is most advanced on the regional or continental level. Economic blocs such as the EU (European Union) and ASEAN (Association of South-East Asian Nations) are increasingly concerned with political issues. The EU is the most politically integrated of the regional economic blocs, with its Commission, Parliaments and other institutions. The impact of this is most felt in the cities where the institutions are based. Brussels, which hosts both the Commission and one of the Parliaments (the other one being in Strasbourg), is very much an international city with a large expatriate European population. In this sense it is a politically globalised city. New York, as the headquarters of the United Nations General Assembly, Secretariat and Security Council, also has a large number of expatriate employees. However, as it has a much greater population than Brussels, the impact is not as great.

iii) Cultural globalisation
The domination of one culture over another is as old as history itself. The imposition of a certain type of architecture and urban form by one people on another has been particularly marked when empires have expanded and conquered vast swathes of new territories. The ancient Greeks and Romans in the Mediterranean and beyond, the Arabs and the Ottomans in the Middle East and North Africa and the Spanish in Latin America all are clear examples of cultures that spread their urban designs into subjucated territories. The British Empire was the first, however, to have an almost global impact on urban planning and architecture. Many of the commercial buildings constructed in the late nineteenth and early twentieth centuries in cities as far flung as Montreal, Melbourne, Singapore, Johannesburg, Bombay (Mumbai) and Nairobi would not look out of place in central Manchester or Birmingham. The use of the grid-pattern layout, imposed by British town planners, is also common to many Commonwealth cities.

In the past few decades, cultural globalisation has taken place at a much greater speed. The end of the Cold War led to the global economic and political domination of capitalism and the USA. This has in turn led to a growing cultural domination of the world by the USA, through the media and entertainment as well as through advertising and other activities carried out by TNCs. In this context, Rizler's concept of the McDonaldisation process, mentioned above, can be regarded as a form of cultural globalisation as much as an economic one. The creation of Disneyland theme parks in Japan and France is another manifestation of cultural globalisation, as is the tendency of US productions to dominate both cinema and television in most parts of the world.

The symbolic 'Downtown' area with its high-rise commercial office blocks, first developed in the cities of the USA, has become the symbol of success, aspired to by any large city the world over. Although some historic buildings such as palaces and places of worship may be preserved within the CBD fabric, the international style of the high-rise office blocks dominates many of the world's city skylines.

The trend towards architectural globalisation is no longer confined to the CBD. Residential areas of cities, which traditionally might have had strong vernacular influences in both building style and materials, are also undergoing a form of globalisation. This is especially marked in rich and middle class suburban areas, and can be seen in countries at all stages of economic development. The type of housing to which so many middle class families aspire is the 'all-American' model, the detached or semi-detached house common in new developments in states such as California and Florida. The style is generally a fusion of the North European and the Spanish colonial, and it is aspired to in many countries as part and parcel of an aspiration to the US lifestyle.

Cities throughout Latin America, eastern and southern Asia and in certain parts of Africa are adopting this 'International American' style in their wealthier suburbs. Singapore provides an excellent example of this. Much of the new housing created in the 1970s and 1980s was high-rise and very functional, and built by the government. Now the new wealthier classes are making a statement by buying larger, more ornate, two- or three-storey houses on privately built estates.

It is quite surprising to see this trend taking a more extreme form in cities within the People's Republic of China. Wu (2003) carried out a study of the development of this globalised housing in Beijing. This development has been stimulated by both the growth of a wealthy middle class and by the urban changes in preparation for the 2008 Olympic Games. There have been over a dozen new townhouse suburban developments in the northern suburbs of Beijing (on the side of the city where the main Olympic Games installations are being built). Some of these housing estates have adopted the 'International American' style, whereas others have, ironically, elements of the English suburban mock Tudor, similar to what was found in the British economic enclave in Shanghai prior to the Second World War. (Similar housing estates are being built in this style in Shanghai too.) Not only is the housing in what Wu calls a style of 'imagined globalisation', but the estates are given Western names associated with wealth, status and success, such as 'Orange County', 'Cambridge', 'Foreign Villa' and 'Times Manor'.

3 High-tech Cities: Past, Present and Future

The high-tech computer era started around the 1960s when the new technologies developed in the Second World War were being applied to peaceful purposes. The electronics, computer and aerospace industries were replacing the old heavy industries as the keys to world economic and political power. The USA, the old USSR, the UK, France and Japan were the leading innovators in these new industries. Very often these activities were given locations away from the old industrial centres. In the case of the USA, the cities of the west or 'Sunbelt' took on the role of centres of new industries replacing those of the north-east or 'Rustbelt'. In Britain the M4 corridor developed as one of the major arteries of these new activities and in France it was the south that attracted a significant part of the high-tech industry, rather than the traditionally industrial north.

Two types of high-tech settlements can be identified: the **technopolis** which is a whole town or city given over to, or dominated by these new industries, and the **technopole** which is a suburb or zone where high-tech industries are concentrated but is located within a bigger settlement. Two high-tech settlements will be examined in more detail here: Sophia-Antipolis in southern France, one of the earliest, and Cyberjaya in Malaysia, one of the newest.

CASE STUDY: SOPHIA-ANTIPOLIS, SOUTHERN FRANCE

Sophia-Antipolis (see Figure 9.1) was the brainchild of the French academic Pierre Laffitte who had the vision of building a science and research park in the south of France. The town, which takes its name from the Greek word for knowledge and the concept of a low-density 'non-city', was founded in 1972 and went into operation in 1974. It is located in the hills above Cannes in the Alpes Maritimes *département* of the Côte D'Azur.

As with so many subsequent high-tech settlements, it is set within a very attractive environment. There are 150 hectares of green recreational spaces within Sophia and a green belt of some 1500 hectares surrounding it, much of which is Mediterranean pine forest, including the three regional forest parks of La Valmasque, Sartoux and La Brague. The town is also well equipped with sports facilities such as golf courses and tennis courts, regarded as necessary to attract the people who work in the high-tech sector. Even more important are the large colleges with state-of-the-art technology for the children of these employees.

Sophia-Antipolis is centred on the CBD of Haute Sartoux and Garbejare and is divided up into eight communities or neighbourhoods. It is well positioned in terms of communications, as the A8 Autoroute passes through its southern suburbs, it is close to both Cannes and Nice international airports and has its own heliport.

Since its inception Sophia has attracted a wide range of industries. In the 1970s, these included IBM, the French National School of Mines, Air France's first computerised reservations unit and the French Petroleum Institute. In the 1980s, an even wider range of activities came to Sophia, including France

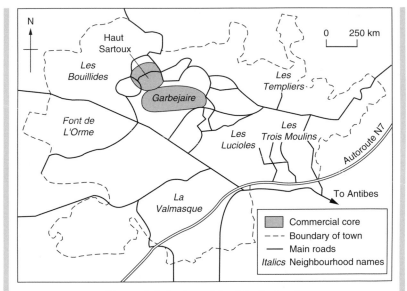

Figure 9.1 A plan of Sophia-Antipolis

Telecom, numerous international firms, the research laboratories of computer companies, and pharmaceutical companies such as the Wellcome Foundation. In the 1990s the high-tech settlement expanded more rapidly with 259 new enterprises locating there in the first half of the decade and a further 133 in the second half. Today there are over 1230 firms and institutes at Sophia and some of its most important activities are now biotechnology, semi-conductor manufacture, software development and travel and tourism companies. The biggest planned extension of the urban area is due to take place within the next decade, when Sophia 2, will be built to the north of the present settlement, almost doubling the size of the technopolis.

CASE STUDY: CYBERJAYA, MALAYSIA

Dr Mahatir Mohammed, Malaysia's long-standing prime minister who retired in 2003, had the vision of making his country an MEDC by the year 2020. Part of this development process was to embark on a number of mega-projects, including the building of Cyberjaya, a high-tech Garden City close to the capital, Kuala Lumpur. Cyberjaya is in fact the nucleus of what the Malaysians call the Multimedia Super Corridor (MSC) which stretches from the ultra-modern international airport to Kuala Lumpur, and also includes the new administrative Garden City of Putrajaya, to which the ministries are being located.

The city was launched in 1999 and is being constructed in two phases of roughly equal size which will eventually cover over 2900 hectares. The population will eventually be 120 000. By 2002, 105 Malaysian companies and 90 foreign firms had already located in Cyberjaya. As with so many modern planned settlements, the city is low density and is subdivided into functional zones. Cyberjaya's three main land-use areas are: the Enterprise Zone, where the main high-tech and other industries are located; the Commercial Zone, where the main shops, hotels and financial services are located; and the Residential Zone, which has a wide variety of housing style and takes up the majority of the land. Broadband Internet technology is linked to every building in the city, which therefore gives it 'smart offices', 'smart schools' and 'smart homes'.

At the centre of all the high technology is the CCC (City Command Centre) that monitors all the main services in Cyberjaya. These include the advanced traffic management, the utilities management (water, electricity and other public utilities) and community services management (e.g. healthcare, schools, rubbish disposal). The CCC can be accessed through telephones and IVR (interactive voice response) systems, interactive televisions and personal computer systems, and mobile data terminals and kiosks in public places.

As with the new administrative capital, Cyberjaya is being landscaped in order to give it the appearance of a resort town. The 300 hectares of recreational park close to the centre of this technopole are the first phase of the Taman Tasik Cyberjaya (the Cyberjaya Lake Garden), which will retain a wide biodiversity of native Malaysian flora and fauna.

4 Sustainability and Cities

a) The meaning of sustainability

The term 'sustainability' is currently overused and misused by politicians and academics alike. Most people have some idea of what it means, but as Fowke and Prasad (1996) pointed out there are at least 80 different definitions of 'sustainable development'. The general concept of sustainability, however, assumes that changes that are taking place today in order to satisfy the needs of the present generations should not compromise the needs of future generations. In the context of the city, this puts a heavy weight on the shoulders of planners and administrators to make decisions and changes that are not going to leave urban areas difficult or even impossible to inhabit in the future.

Many different factors enter into any debate on cities and their sustainability:

*   The problem of size, whether this be in terms of the total area of the city or conurbation or the total population. Is there an ideal or optimum size for a city? Can large cities be allowed to continue to grow?
*   The problems of congestion, overcrowding and traffic growth, and the effects they have on movement of people, accessibility, pollution levels and the urban microclimate.
*   The pressures placed on resources and services, such as water supplies, energy supplies, and land for residential, retail and industrial developments.
*   The additional demands for housing and recreational facilities in the countryside beyond the city thereby involving the rural fringes in any sustainability equation.

b) Progress in achieving sustainable cities

MEDCs, in general, have much better resources and can afford to adopt greener policies towards their towns and cities. However, the political will is necessary and there are huge variations between MEDCs and their planning policies. Politicians tend to work on the basis of short-term time scales, whereas sustainability is a matter of the long-term perspective. Countries such as the Netherlands, Sweden and Denmark have had a long tradition of making their towns and cities into attractive living environments with efficient public transport and low pollution levels. Old industrialised countries such as the UK and Germany have had more of a problem of converting their industrial cities with poor housing and derelict, obsolete factories into cleaner, greener post-industrial settlements.

With the collapse of Communism in Eastern Europe in the early 1990s, the task has been even more difficult, as the old political system was so emphatically gauged to increasing industrial production at all costs and placed little emphasis on safeguarding the environment. Cities in countries such as Poland, Romania and the Czech Republic still have a long way to go to reach the sustainability that has been achieved in much of north-west Europe. Many cities within Mediterranean Europe were in past decades chaotic in their planning and development because of high demographic and economic growth rates as well as high pollution levels. As population growth has levelled off, certain Mediterranean cities have been able to work towards greater sustainability, especially when they have had enlightened mayors or have been the setting of important world events. Barcelona, when it became the host of the Olympic Games in 1992, underwent huge changes in clearing up old industrial sites and polluted coasts and extended its public transport network; since then it has become one of Europe's most frequently visited tourist spots. Rome, under the two centre-left mayors, Rutelli and Veltroni, from the late 1990s, and with the stimulus of the Millennium Jubilee in 2000 has undergone a huge transformation with restoration of hundreds of buildings, the growth of cleaner, better public transport and the improvement of many of its public services.

Within LEDCs, there are also wide-ranging contrasts between the levels of sustainability achieved, but by and large, the biggest and fastest growing conurbations are those which are most held back in their quests for sustainability. Singapore, which is as rich as, if not richer than most European cities, shows what can be done in a location very close to the equator. It has vast areas of greenery, high standards of public housing, health provision and education, good, cheap and efficient public transport, and private traffic discouraged from the CBD. Singapore, however, as it is an NIC, is not representative of the tropical world as a whole. Elsewhere megacities struggle to cope with heavy inward migration, poverty, shanty town settlements and the whole range of interrelated problems associated with LEDC cities. Whether it be Cairo or Kinshasa in Africa, Bogotá or Belém in Latin America, or Mumbai or Manila in Asia, the problems are much the same and any form of sustainability may not be reached for several decades. Yet there are oases of hope within LEDCs. Curitiba in Brazil has become a model of urban sustainability within poorer countries (its well-organised public transport system was examined in Chapter 6).

The cities that are at present the least sustainable are those that have large demographic increases from refugees and or have been ravaged by war. Examples of cities where recent wars have destroyed infrastructure and essential services, making them unable to support their existing populations at an acceptable level, include Mogadishu in Somalia (examined in the previous chapter), Freetown in Sierra Leone, Monrovia in Liberia, Grozny in Chechnya (Russian Federation), Kisingani in the Congo, and Baghdad and Basra in Iraq.

CASE STUDY: LONDON'S URBAN 'METABOLISM'

Cities, like humans, have a type of metabolism which involves the intake of energy and other resources and the output of waste products. At present, London is far from sustainable; it has 12% of the country's population and its **ecological footprint** (i.e. the area from which it draws its main resources to support its needs) stretches over an area 125 times larger than Greater London itself. The built-up area is 158 000 hectares (ha) in extent, yet the area of farmland required to feed its people is some 8.4 million ha, the forest area needed to provide the people with their need for timber is some 768 000 ha, and the area required to produce its fuel needs is some 10.5 million ha.

5 Future Cities

Whereas rural areas are generally associated with tradition and conservatism, cities are all about change. As John Rennie Short (1991) puts it:

> 'The city is the metaphor for social change, an icon of the present at the edge of transformation of the past into the future. Attitudes about the city reflect attitudes about the future.'

In several parts of this book the current situation in the USA, and particularly the Los Angeles area, has been examined as an indicator of the future patterns of urbanism. The decentralisation of functions, the growth of ever-sprawling suburbs, the emergence of Edge Cities, Off-the-Edge Cities, Edgeless Cities and Exopolis, together with the polarisation of rich and poor, are at present the future models being provided by the USA. By contrast, the more sustainable concepts of urban growth, which would make cities more liveable, are coming from Europe and other parts of the world. Technological change is the greatest unknown quantity in the future. At present, technopoles in both the USA and Europe may point to the future, along with the 'wired cities' of Asia, such as Singapore, Cyberjaya and Shanghai. But just as current technologies, particularly within the realms of information technology, could not have been predicted 50 years ago, it is impossible to know what could transform cities in 50 years' time.

Already technological changes have had many influences on location, cities and geographical patterns, some of the most important of which are:

- the changes in accessibility: telecommunications can in many ways be a substitute for physical movement
- mobility: people can participate in activities far beyond their local area
- technology: activities are now increasingly scattered between different locations; no longer do people just work from the office, but also from home, the hotel room, from planes, trains and cars. Also, settlements have within them segregated, affluent enclaves of the technology-rich in contrast to wider areas of the technology-poor; technology can create wider social polarisation in cities.

Although it is impossible to predict the future, Pacione (2001) puts forward seven different scenarios for how cities may develop in the future, and indeed all seven of these could co-exist alongside each other in the future world:

- the **green city**, following the model of Ebenezer Howard and his Garden Cities into a future where ecologists and the green movement have their way and develop sustainable cities in harmony with nature

- the **dispersed city**, also following the move to a more sustainable future, would see urban functions spread into small communities which could be more self-supporting through recycling
- the **compact city**, following the Le Corbusier model and avoiding urban sprawl, this would be a high-density, high-rise city which would be energy efficient, reducing the need for commuting
- the **network city**, which already exists in those world cities that were discussed above as being part of a global business network; this type of city may also be a close-knit network of urban areas within a country, as in Randstad Holland
- the **informational city**, following the model of high-tech cities of today; in the future these post-Industrial cities increasingly will be able to replace actual flows of people and traffic with telematic flows of information, an environmentally friendly situation
- the **virtual city**, which takes telematics one stage further, into the realms of what is now science fiction, and this polarises opinions; as Pacione puts it: 'For some "cyber-utopians" the virtual city represents a future urban environment liberated from the constraints of place-bound interaction. For others the virtual city heralds a dystopian urban future characterised by the kind of social disintegration portrayed in films such as *Blade Runner* and *Judge Dredd*.'

Questions

1. Discuss the differences between the concepts of **world cities** and global cities in the context of the changing world economy of the last few decades.
2. Assess the impact of **globalisation** in its many forms upon cities in the contemporary world.
3. How can cities and their development become **sustainable**?
4. Consider, in the light of present experience, the different ways in which cities may evolve in the future.

Bibliography

Aliaj, B. *et al.*, 2003. *Tirana: The Challenge of Urban Development.* Tirana: SEDA.

Besana, R. *et al.*, 2002. *Metaphisica Costruita.* Milan: Touring Club Italiano.

Bridge, G. and Watson, S., 2000. *A Companion to the City.* Oxford: Blackwell.

Burke, G., 1971. *Towns in the Making.* London: Edward Arnold.

Burtenshaw, D., 1983. *Cities and Towns.* London: Bell & Hyman.

Burtenshaw, D., *et al.*, 1991. *The European City: A Western Perspective.* London: David Fulton.

Carter, H., 1990. *Urban and Rural Settlements.* London: Longman.

Castells, M. and Hall, P., 1994. *Technopoles of the World.* New York: Routledge.

Cloke, P., Crang, P. and Goodwin, M. eds, 1999. *Introducing Human Geographies.* London: Arnold.

Cosh, M., 1981. *An Historical Walk Through Barnsbury.* London: Islington Archaeology and Historical Society.

Daniel, P. and Hopkinson, M., 1989. *The Geography of Settlement*, 2nd edition. Harlow: Oliver & Boyd.

Dear, M. and Flusty, S., 2001. *The Spaces of Postmodernity: Readings in Human Geography.* Oxford: Blackwell.

Downey, J. and McGuigan, J. eds, 1999. *Technocities.* London: Sage.

Drake, G. and Lee, C., 2000. *The Urban Challenge.* London: Hodder & Stoughton.

Faulconbridge, J., 2004. London and Frankfurt in Europe's evolving financial centre network. *Area* Vol. 36, No. 3. London: RGS-IBG.

Fowke, R. and Prasad, D. 1966. Sustainable development, cities and local government. *The Australian Planner* Vol. 33.

Friedmann, J. and Wolff, K., 1982. Word city formation for research and action. *International Journal of Urban and Regional Research* Vol. 6, No. 3.

Fukuyama, F., 2004. *State Building: Governance and the World Order in the Twenty-First Century.* London: Profile Books.

Garreau, J., 1991. *The Edge City: Life on the New Frontier.* New York: Doubleday.

Geddes, P., 1915. *Cities in Evolution.* London: Williams & Norgate.

Glass, R., 1963. *London: Aspects of Change.* London: Centre for Urban Studies.

Goodman, D. and Chant, C. eds, 1999. *European Cities and Technology.* London: Routledge.

Gratz, R., 1998. *Cities Back from the Edge.* New York: John Wiley.

Hall, P., 1966. *The World Cities.* London: Weidenfeld & Nicolson.

Hall, T., 1998. *Urban Geography.* London: Routledge.

Hamnett, C., 2003. *Unequal City – London in the Global Arena.* London: Routledge.

Hannigan, J., 1999. *Fantasy City.* London: Routledge.

Harvey, D., 1989. *The Condition of Postmodernity.* Oxford: Blackwell.

Hill, M., 1999. *Advanced Geography Case Studies.* London: Hodder & Stoughton.

Hill, M., 2003. *Access to Geography: Rural Settlement and the Urban Impact on the Countryside.* London: Hodder & Stoughton.

Hoskins, W.G., 1955. *The Making of the English Landscape.* London: Hodder & Stoughton.

Jenkins, J. ed., 2002. *Remaking the Landscape.* London: Profile Books.

Kostof, S., 1992. *The City Assembled.* London: Thames & Hudson.

Lang, R., 2003. *Edgeless Cities: Exploring the Elusive Metropolis.* Washington DC: Brookings Institution Press.

Lenon, B., 1993. *London in the 1990s.* Sheffield: The Geographical Association.

Lowry, J., 1975. *World City Growth.* London: Arnold.

Lycett Green, C., 1996. *England: Travels Through An Unwrecked Landscape.* London: Pavilion Books.

Marcuse, P. and van Kempen, R., 2000. *Globalizing Cities: A New Spatial Order?* Oxford: Blackwell.

Martinelli, R. and Nuti, L. eds, 1978. *Le Città di Fondazione.* Lucca: CISCU-Marsilio Editori.

Minca, C. ed., 2001. *Postmodern Geography: Theory and Praxis.* Oxford: Blackwell.

Morris, J., 1986. *Among the Cities.* Harmondsworth: Penguin.

Ohmae, K., 1995. *The End of the Nation State.* London: Macmillan.

Pacione, M., 2001. *Urban Geography: A Global Perspective.* London: Routledge.

Pretty, J., 1998. *The Living Land.* London: Earthscan.

Raban, J., 1974. *Soft City.* London: Harvill Press.

Risebero, B., 1992. *Fantastic Form: Architecture and Planning Today.* London: Herbert Press.

Rykwert, J., 2000. *The Seduction of Place: The City in the 21st Century.* London: Weidenfield & Nicolson.

Savitch, V., 1988. *Politics and Planning in New York, Paris and London.* Princeton NJ: Princeton University Press.

Short, J.R., 1991. *Imagined Country: Society, Culture and Environment.* London: Routledge.

Short, J.R. and Kim, Y.-H., 1999. *Globalization and the City.* Harlow: Longman.

Soja, E., 2000. *Postmetropolis: Critical Studies of Cities and Regions.* Oxford: Blackwell.

Speak, J. and Fox, V., 2002. *Regenerating City Centres.* Sheffield: Geographical Association.

Taylor, P.J., 2001. Urban hinterworlds: geographies of corporate service provision under conditions of contemporary globalisation. *Geography* Vol. 86, No. 1. Sheffield: Geographical Association.

Thrift, N., 1996. *Spatial Formations.* London: Sage.

Towers, G., 1995. *Building Democracy: Community Architecture in the Inner Cities.* London: UCL Press.

UN-Habitat, 2004. *The State of the World's Cities 2004/2005.* London: Earthscan.

Valentine, G., 2001. *Social Geographies: Space and Society.* Harlow: Pearson Education.

Wu, F., 2004. Transplanting cityscapes: the use of imagined globalisation in housing commodification in Beijing. *Area* Vol. 36, No. 3. London: RGS-IBG.

Websites

* All UK universities that have geography departments have useful websites. These, along with departments in other countries, can be accessed through the IGU (type IGU into a search engine and follow instructions or www.igu-net.org).
* The Royal Geographical Society. www.rgs.org
* The Geographical Association. www.geography.org.uk
* Urbis in Manchester: a museum/exhibition of city life throughout the world. www.urbis.org.uk

Index